Lecture Notes in Mathematics

Edited by A. Dold and B. Eckmann

715

Inder Bir S. Passi

Group Rings and Their Augmentation Ideals

Springer-Verlag
Berlin Heidelberg New York 1979

Author

Inder Bir S. Passi
Department of Mathematics
Kurukshetra University
Kurukshetra (Haryana)
India 132119

AMS Subject Classifications (1970): 16A26, 20C05

ISBN 3-540-09254-4 Springer-Verlag Berlin Heidelberg New York
ISBN 0-387-09254-4 Springer-Verlag New York Heidelberg Berlin

Printing and binding: Beltz Offsetdruck, Hemsbach/Bergstr.
2141/3140-543210

INTRODUCTION

The purpose of these Notes is to report on some aspects of the augmentation ideal of a group ring.

Every two-sided ideal in a group ring determines a normal subgroup. One of the oldest and most challenging problems in group rings is the identification of the normal subgroups, called dimension subgroups, determined by the powers of the augmentation ideal of a group ring. The first five Chapters of these Notes are devoted to this problem.

Dimension subgroups were first studied by Magnus [48] who proved that, for free groups, the integral dimension subgroup series coincides with the lower central series. More than thirty years later Rips [78] demonstrated that Magnus' theorem cannot be generalized to arbitrary groups.

Sandling ([79], [80]) initiated the study of dimension subgroups over arbitrary coefficient rings and of Lie dimension subgroups. Both the associative and the Lie powers [79] of an augmentation ideal are polynomial ideals [59]. We, therefore, begin in Chapter I with the study of polynomial ideals. The main result proved in Chapter II is that for the study of (Lie) dimension subgroups over arbitrary coefficient rings it is enough to restrict to integral and modular coefficients. Chapter III is an exposition of a portion of a fundamental paper of Hartley [27]. Our aim here is to show that Hartley's theory provides a frame work in which it is possible to unify the approaches of Jennings ([35], [36]) and Lazard [38] for the study of augmentation powers. Dimension subgroups over fields have been studied by Bovdi [7], Hall [26], Jennings ([35], [36]), Lazard [38], Sandling [80] and Zassenhaus [99]. We calculate these subgroups in Chapter IV. We study in Chapter V polynomial maps on groups and calculate integral dimension subgroups in certain cases. At the present time, apart from some special classes of groups, complete description of only the first five integral dimension subgroups is known and the general problem remains as challenging as ever.

Chapters VI and VII are devoted to the study of the intersection of the powers of an augmentation ideal. Lichtman [39] has recently characterized group theoretically the residual nilpotence of the augmentation ideal (over integers). A reasonable description of the intersection of the powers of an augmentation ideal, however, is still not known.

The associated graded ring arising from the filtration of a group ring given by the powers of its augmentation ideal is the subject of study in Chapter VIII. This is an object which is extremely rich in algebraic structure. As for dimension subgroups, the structure of the associated graded ring can be nicely described when the coefficients are in a field (Quillen [77]). However, the general case remains little understood.

I would like to express my sincere thanks to my teacher D. Rees who introduced me to group rings. I am indebted to Brian Hartley and Robert Sandling for critically reading parts of these Notes and making several helpful comments. The writing of these Notes was undertaken during the year 1973-74 when I was visiting the University of Alberta. The completion of these Notes is the result of my stay at the University of Warwick during the 1978 Symposium on Infinite Groups and Group Rings. The hospitality of the University of Alberta and the University of Warwick is gratefully acknowledged. I am thankful to my colleagues J.N. Mital and L.R. Vermani and research students M. Goyal and S.K. Arora for numerous stimulating discussions. Thanks are also due to K.W. Gruenberg, N.D. Gupta, D.S. Passman and U. Stammbach for their interest in these Notes and the encouragement they have given me.

September, 1978

Kurukshetra University
Kurukshetra (Haryana)
India 132119

Current address: Centre for Advanced Study
in Mathematics
Panjab University
Chandigarh-160014
India

CONTENTS

CHAPTER I

POLYNOMIAL IDEALS IN GROUP RINGS

Every element of an integral group ring of a free group defines, via substitution, an ideal in an arbitrary group ring. Such ideals are here called polynomial ideals. Both the associative and the Lie powers of an augmentation ideal are polynomial ideals. The main result, Theorem 1.12, of this Chapter relates polynomial ideals in arbitrary group rings to those in integral and modular group rings.

1. POLYNOMIAL IDEALS

Let G be a group and R a commutative ring with identity. We denote by $R(G)$ the group ring of G over R ; i.e. $R(G)$ is the free R-module on the set G with multiplication defined distributively using the group multiplication in G . We denote by 1_R the identity of R and by 1_G the identity of G . However, we will be dropping the suffix R or G if there is no danger of confusion. We regard R as a subring of $R(G)$ by identifying $r \in R$ with $r1_G \in R(G)$.

Let F be a free group on the free generators x_i $(i \in I)$, say, and $Z(F)$ be its integral group ring (Z denotes the ring of rational integers). Then every homomorphism $\alpha : F \to G$ induces a ring homomorphism $\alpha^*: Z(F) \to R(G)$, $\alpha^*(\Sigma n\,\omega) = \sum_{n \in Z, \omega \in F} (n\,1_R)\alpha(\omega)$. If $f(x) \in Z(F)$, we denote by $A_{f,R}(G)$ the two-sided ideal of $R(G)$ generated by the elements $\alpha^*(f(x))$, $\alpha \in \mathrm{Hom}(F,G)$, the set of homomorphisms from F to G . In other words $A_{f,R}(G)$ is the ideal generated by the values of $f(x)$ in $R(G)$ as the elements of G are substituted for the free generators x_i's .

1.1 Definition. An ideal I of $R(G)$ is called a polynomial ideal if $I = A_{f,R}(G)$ for some $f \in Z(F)$, F a free group.

1.2 Examples. (1) The R-homomorphism $\varepsilon : R(G) \to R$ induced by mapping every element of G into 1_R is called the unit augmentation. We denote the kernel of ε by $\Delta_R(G)$. It is a two-sided ideal of $R(G)$ and is called the augmentation ideal of $R(G)$. As an R-module, $\Delta_R(G)$ is a free module with the elements $g-1$ $(1 \neq g \in G)$ as a basis.

The augmentation ideal $\Delta_R(G)$ is a polynomial ideal. For, $\Delta_R(G)$ is generated as an R-module by the elements $g - 1_R$, $g \in G$, i.e. by the values of the polynomial $x - 1$. In fact, all associative powers $\Delta_R^i(G)$, $i \geq 1$, of $\Delta_R(G)$ are polynomial ideals:

$$\Delta_R^i(G) = A_{f,R}(G) \text{ , where } f = (x_1-1)(x_2-1)\dots(x_i-1) \text{ .}$$

(2) The Lie powers $\Delta_R^{(i)}(G)$ of the augmentation ideal are defined inductively as follows:

$$\Delta_R^{(1)}(G) = \Delta_R(G) \text{ , } \Delta_R^{(i+1)}(G) = [\Delta_R(G), \Delta_R^{(i)}(G)]R(G) \text{ ,}$$

where $[M,N]$ denotes the R-submodule of $R(G)$ generated by $[m,n] = mn - nm$, $m \in M$, $n \in N$ and , for $K \subseteq R(G)$, $K \cdot R(G)$ denotes the right ideal generated by K in $R(G)$ (Similarly $R(G) \cdot K$ will denote the left ideal generated by K). Since

$$g[\alpha,\beta] = [g\alpha g^{-1}, g\beta g^{-1}]g \text{ , } g \in G \text{ , } \alpha,\beta \in R(G) \text{ ,}$$

induction shows that the right ideal $\Delta_R^{(i)}(G)$ is a two-sided ideal of $R(G)$ for all $i \geq 1$. We shall prove that $\Delta_R^{(i)}(G)$ is a polynomial ideal for every $i \geq 1$ (Corollary 1.9).

1.3 Definition. A free polynomial is an element $f(x)$ of the integral group ring $Z(F)$ of a free group F .

The following Proposition is obvious.

1.4 Proposition. Let $f(x)$ be a free polynomial. Then $f(x)$ defines a polynomial ideal $A_{f,R}(G)$ in every group ring $R(G)$. Further, if $\alpha^*: R(G) \to S(H)$ is a ring homomorphism induced by a group homomorphism $\theta : G \to H$ and a ring homomorphism $\varphi : R \to S$, then

$$\alpha^*(A_{f,R}(G)) \subseteq A_{f,S}(H) \text{ .}$$

[It is assumed here that $\varphi(1_R) = 1_S$.]

Proposition 1.4 shows that a free polynomial defines an invariant ideal in the sense of K.T. Chen [9].

1.5 Proposition. Let $A = A_{f_1,R}(G)$ and $B = A_{f_2,R}(G)$ be two polynomial ideals of $R(G)$ defined by the free polynomials $f_1(x_1,x_2,\dots,x_m)$ and $f_2(x_1,x_2,\dots,x_n)$ respectively. Then $A+B$ and AB are also polynomial ideals provided both f_1 and f_2 have content zero (i.e. the sum of the coefficients is zero.)

Proof. Let $f(x_1,x_2,\dots,x_{m+n}) = f_1(x_1,x_2,\dots,x_m) + f_2(x_{m+1},x_{m+2},\dots,x_{m+n})$. Since both f_1 and f_2 have content zero, we have, in $R(G)$, $f_1(1_G,1_G,\dots,1_G) = 0$ and $f_2(1_G,1_G,\dots,1_G) = 0$ and it is clear that $A_{f,R}(G) = A+B$.

Let $f(x_1,x_2,\dots,x_{m+n}) = f_1(x_1,x_2,\dots,x_m)f_2(x_{m+1},x_{m+2},\dots,x_{m+n})$.

Then $A_{f,R}(G) \subseteq A_{f_1,R}(G)A_{f_2,R}(G)$. The reverse inclusion follows from the observation

$$f_1(u_1,u_2,\ldots,u_m)u = uf_1(u_1^u,u_2^u,\ldots,u_m^u) ,$$

where u_i's and $u \in G$ and $u_i^u = u^{-1}u_iu$. Thus $AB = A_{f,R}(G)$.

Let $w = w(x_1,x_2,\ldots,x_n)$ be a word in $x_1,x_2,\ldots,x_n$; i.e. an element of the free group F generated by $x_1,x_2,\ldots,x_n$. For a group G , we denote by $w(G)$ the <u>verbal subgroup</u> of G defined by w , i.e. the subgroup generated by the elements $w(g_1,g_2,\ldots,g_n)$, $g_i \in G$. Since

$$w(g_1,g_2,\ldots,g_n)^g = w(g_1^g,g_2^g,\ldots,g_n^g) \quad , g \in G \text{ and } g_i\text{'s} \in G ,$$

$w(G)$ is a normal subgroup of G .

<u>For a normal subgroup</u> N <u>of</u> G , <u>let</u> $\Delta_R(G,N)$ <u>denote the kernel of the natural epimorphism</u> $R(G) \to R(G/N)$ <u>induced by</u> $G \to G/N$. It may be observed that $\Delta_R(G,N)$ is the two-sided ideal of $R(G)$ generated by $\Delta_R(N)$. Moreover, because of the normality of N ,

$$\Delta_R(G,N) = \Delta_R(N)\cdot R(G) = R(G)\cdot\Delta_R(N) .$$

If $x,y \in G$, then we have the following equations in $R(G)$:

$$xy - 1 = (x-1) + (y-1) + (x-1)(y-1)$$

$$x^{-1} - 1 = -(x-1)x^{-1} .$$

Thus, if N is generated as a subgroup by a subset (h_λ) , then every element of $\Delta_R(G,N)$ can be written as $\Sigma(h_\lambda-1)\alpha_\lambda$, where $\alpha_\lambda \in R(G)$.

1.6 <u>Proposition</u>. <u>For every word</u> w , $\Delta_R(G,w(G))$ <u>is a polynomial ideal</u>.

<u>Proof</u>. Let $f(x_1,x_2,\ldots,x_n) = w(x_1,x_2,\ldots,x_n) - 1$. Then

$$A_{f,R}(G) = \Delta_R(G,w(G)) .$$

For a group G we denote by $\gamma_i(G)$, $i \geq 1$, the i-th term in the <u>lower central series</u> of G . The subgroups $\gamma_i(G)$ are defined inductively by

$$\gamma_1(G) = G , \quad \gamma_{i+1}(G) = (G,\gamma_i(G)), \quad i \geq 1 ,$$

where (H,K) denotes the subgroup generated by the commutators

$$(h,k) = h^{-1}k^{-1}hk , \quad h \in H , \ k \in K .$$

It may be recalled that the subgroups $\gamma_i(G)$ are verbal subgroups. If $w_i = (\ldots((x_1,x_2),x_3),\ldots,x_i)$, then $\gamma_i(G) = w_i(G)$ (for example, see

[25], p. 150).

The following Proposition gives some elementary properties of the Lie powers $\Delta_R^{(i)}(G)$.

1.7 Proposition.

(i) $\Delta_R(G,\gamma_n(G)) \subseteq \Delta_R^{(n)}(G)$ for all $n \geq 1$

(ii) $[\Delta_R^{(m)}(G),\Delta_R^{(n)}(G)] \subseteq \Delta_R^{(m+n)}(G)$ for all $m,n \geq 1$.

(iii) $\Delta_R^{(m)}(G)\Delta_R^{(n)}(G) \subseteq \Delta_R^{(m+n-1)}(G)$ for all $m,n \geq 1$.

Proof. (i) We proceed by induction on n , the statement being clear for $n = 1$. Let $g \in G$, $h \in \gamma_{n-1}(G)$ and assume that (i) holds for $n-1$ where n is an integer ≥ 2 . Then

$$\begin{aligned}(g,h) - 1 &= g^{-1}h^{-1}gh - 1\\ &= g^{-1}h^{-1}(gh-hg)\\ &= g^{-1}h^{-1}[g-1,h-1] \in \Delta_R^{(n)}(G) ,\end{aligned}$$

since $h - 1 \in \Delta_R^{(n-1)}(G)$ by induction hypothesis. Since $\Delta_R(G,\gamma_n(G))$ is generated by elements of the type $(g,h) - 1$, $g \in G$, $h \in \gamma_{n-1}(G)$, (i) is proved.

(ii) and (iii) both are trivially true for $m = 1$ and all $n \geq 1$. Suppose that both hold for some $m \geq 1$ and all $n \geq 1$. Then the identity

$$[ab,c] = a[b,c] + [a,c]b \qquad (*)$$

in $R(G)$ shows that

$$\begin{aligned}&[\Delta_R^{(m+1)}(G),\Delta_R^{(n)}(G)] = [[\Delta_R(G),\Delta_R^{(m)}(G)]R(G),\Delta_R^{(n)}(G)]\\ &\subseteq [\Delta_R(G),\Delta_R^{(m)}(G)][R(G),\Delta_R^{(n)}(G)] + [[\Delta_R(G),\Delta_R^{(m)}(G)],\Delta_R^{(n)}(G)]R(G)\\ &\subseteq [\Delta_R(G),\Delta_R^{(m)}(G)]\Delta_R^{(n+1)}(G) + [[\Delta_R(G),\Delta_R^{(m)}(G)],\Delta_R^{(n)}(G)]R(G) .\end{aligned}$$

The identity (*) and the induction hypothesis show that

$$[\Delta_R(G),\Delta_R^{(m)}(G)]\Delta_R^{(n+1)}(G) \subseteq \Delta_R^{(m+n+1)}(G) .$$

The Jacobi identity

$$[[a,b],c] + [[b,c],a] + [[c,a],b] = 0$$

in $R(G)$ and the induction hypothesis show that

$$[[\Delta_R(G),\Delta_R^{(m)}(G)],\ \Delta_R^{(n)}(G)]R(G) \subseteq \Delta_R^{(m+n+1)}(G) .$$

Hence

$$[\Delta_R^{(m+1)}(G),\Delta_R^{(n)}(G) \subseteq \Delta_R^{(m+n+1)}(G) \quad \text{for all} \quad n \geq 1 .$$

Finally the identity (*) and the induction hypothesis show that

$$[\Delta_R(G),\Delta_R^{(m)}(G)]\Delta_R^{(n)}(G) \subseteq \Delta_R^{(n+m)}(G)$$

and therefore

$$\Delta_R^{(m+1)}(G)\cdot\Delta_R^{(n)}(G) \subseteq \Delta_R^{(n+m)}(G)$$

for all $n \geq 1$. This completes the induction and both (ii) and (iii) are established.

We next give Sandling's formula for $\Delta_R^{(n)}(G)$.

1.8 Theorem [79]. Let G be a group, R a commutative ring with identity. Then

$$\Delta_R^{(n)}(G) = \Delta_R(G,\gamma_n(G)) + \Sigma\Pi_j\Delta_R(G,\gamma_{n_j}(G)) ,$$

where the sum is over all partitions n_j , $n \geq n_j > 1$, for which $\Sigma(n_j-1) = n - 1$.

Proof. Proposition 1.7 (i) and (iii) show that the right hand side is contained in $\Delta_R^{(n)}(G)$. To show that $\Delta_R^{(n)}(G)$ is contained in the right hand side we proceed by induction on n , the assertion being clear for $n = 2$. Let $n > 2$ and assume that the theorem holds for $n - 1$, i.e.

$$\Delta_R^{(n-1)}(G) \subseteq \Delta_R(G,\gamma_{n-1}(G)) + \Sigma\Pi_j\Delta_R(G,\gamma_{n_j}(G)) ,$$

where the sum is over all n_j , $n - 1 \geq n_j > 1$, for which $\Sigma(n_j-1) = n - 2$.

Let $u,v \in G$ and $h \in \gamma_{n-1}(G)$. Then by (*)

$$\begin{aligned}[(h-1)u,v-1] &= (h-1)[u,v-1] + [h-1,v-1]u \\ &= (h-1)vu((u,v)-1) + vh((h,v)-1)u .\end{aligned}$$

The first term belongs to $\Delta_R(G,\gamma_{n-1}(G))\cdot\Delta_R(G,\gamma_2(G))$ and the second term belongs to $\Delta_R(G,\gamma_n(G))$, since $(h,v) \in \gamma_n(G)$. This shows that

$$[\Delta_R(G,\gamma_{n-1}(G)),\Delta_R(G)] \subseteq \Delta_R(G,\gamma_n(G)) + \Delta_R(G,\gamma_{n-1}(G))\Delta_R(G,\gamma_2(G)) .$$

It thus suffices to show that for $n - 1 \geq n_j > 1$, $\Sigma(n_j-1) = n - 2$, $[\Pi_j\Delta_R(\gamma_{n_j}(G))\cdot R(G),\Delta_R(G)]$ is contained in

$$\Delta_R(G,\gamma_n(G)) + \Sigma\Pi_i\Delta_R(G,\gamma_{m_i}(G)) ,$$

where the sum is over all m_i , $n \geq m_i > 1$, $\Sigma(m_i-1) = n-1$. By (*) we see that

$$[\Pi_j \Delta_R(\gamma_{n_j}(G))\cdot R(G), \Delta_R(G)]$$

is contained in

$$\Pi_j \Delta_R(\gamma_{n_j}(G))\Delta_R(\gamma_2(G))\cdot R(G) + [\Pi_j \Delta_R(\gamma_{n_j}(G)), \Delta_R(G)]R(G) .$$

The first product is of the desired type by definition. That the second term is of the required type can be seen by a repeated application of (*) and the observation that

$$[\Delta_R(\gamma_i(G)), \Delta_R(G)] \subseteq \Delta_R(G, \gamma_{i+1}(G)) .$$

This completes the induction and Theorem 1.8 is proved.

Since $\gamma_i(G)$, $i \geq 1$, is a verbal subgroup, the ideal $\Delta_R(G, \gamma_i(G))$ is a polynomial ideal (Proposition 1.6). Thus, by Proposition 1.5, we have

1.9 Corollary. The Lie powers $\Delta_R^{(i)}(G)$, $i \geq 1$, are polynomial ideals in $R(G)$.

Let G be a group, R a commutative ring with identity.

1.10 Definition. Let f be a free polynomial. Then a map $\theta : G \to M$, M an R-module, is called an f_R-polynomial map if the linear extension $\theta^*: R(G) \to M$ of θ to $R(G)$ vanishes on $A_{f,R}(G)$, the polynomial ideal determined by f .

[f_R-polynomial maps with $R = Z$ and $f = (x_1-1)(x_2-1)\ldots(x_n-1)$ have been studied in [62]. For further details see Chapter V.]

Let f be a free polynomial. Then the map $\lambda_f: G \to R(G)/A_{f,R}(G)$ given by

$$\lambda_f(g) = g + A_{f,R}(G) , \quad g \in G$$

is evidently an f_R-polynomial map. Note that if $\theta : G \to M$, M an R-module, is an f_R-polynomial map, then $\theta^*: R(G) \to M$ induces an R-homomorphism

$$\bar{\theta}^*: R(G)/A_{f,R}(G) \to M$$

and

$$\theta(g) = \bar{\theta}^*(\lambda_f(g)) , \quad g \in G .$$

In other words, every f_R-polynomial map factors through the f_R-polynomial map λ_f .

For every group G and a commutative ring R with identity, we have a ring homomorphism

$$i_R = i_R(G) : Z(G) \to R(G)$$

given by

$$i_R(\Sigma n_g g) = \Sigma(n_g 1_R)g$$

$g \in G$, $n_g \in Z$.

Given a polynomial ideal $A_{f,R}(G)$, we wish to investigate the inverse image of $A_{f,R}(G)$ in $Z(G)$ under the ring homomorphism $i_R(G)$. One of our main aims in this work is the investigation of the subsets

$$G \cap (1_R + \Delta_R^n(G)) = \{g \in G \mid i_R(g-1) \in \Delta_R^n(G)\}$$

and

$$G \cap (1_R + \Delta_R^{(n)}(G)) = \{g \in G \mid i_R(g-1) \in \Delta_R^{(n)}(G)\} .$$

Since both $\Delta_R^n(G)$ and $\Delta_R^{(n)}(G)$ are polynomial ideals for all $n \geq 1$, the study of the subsets

$$i_R^{-1}(A_{f,R}(G)) = \{z \in Z(G) \mid i_R(z) \in A_{f,R}(G)\}$$

for polynomial ideals helps to unify the investigation of $G \cap (1+\Delta_R^n(G))$ and $G \cap (1+\Delta_R^{(n)}(G))$. This is actually our motivation for introducing polynomial ideals. It turns out that the inverse image $i_R^{-1}(A_{f,R}(G))$ under the ring homomorphism $i_R(G)$ depends only on the characteristic of R (= the least non-negative integer n such that $n1_R = 0$), the behaviour of the elements $p1_R$, p prime, $A_{f,Z}(G)$ and $i^{-1}_{Z/p^eZ}(A_{f,Z/p^eZ}(G))$.

We need the following

1.11 Proposition. Let M be an Abelian group, N an R-module, $\theta : G \to M$ an f_Z-polynomial map, $\varphi : M \to N$ a homomorphism. Then the map $\varphi \circ \theta : G \to N$ is an f_R-polynomial map.

Proof. We first observe that as an R-module the two-sided ideal $A_{f,R}(G)$ is generated by the elements $i_R(g\bar{f})$ where $g \in G$ and $\bar{f} \in Z(G)$ is an f-value. Since

$$(\varphi \circ \theta)^*(i_R(g\bar{f})) = \varphi(\theta^*(g\bar{f})) = 0 ,$$

the result follows (* denotes the linear extension of the map to the group ring).

We now come to the main result of this Chapter.

1.12 <u>Theorem</u> [59]. *Let* $f(x_1,x_2,\dots,x_n)$ *be a free polynomial*, G *a group*, R *a commutative ring with identity. Then*

(i) *if characteristic of* $R = r > 0$,

$$i_R^{-1}(A_{f,R}(G)) = i_{Z/rZ}^{-1}(A_{f,Z/rZ}(G)) = \cap\, i^{-1}_{Z/p^eZ}(A_{f,Z/p^eZ}(G)) ,$$

where the intersection is over all primes p *dividing* r *and* p^e *is the highest power of* p *that divides* r ;

(ii) *if characteristic of* $R = 0$,

$$i_R^{-1}(A_{f,R}(G)) = \sum_{p\in\sigma(R)} \{\tau_p(Z(G) \bmod A_{f,Z}(G)) \cap i^{-1}_{Z/p^eZ}(A_{f,Z/p^eZ}(G))\}$$

where $\sigma(R)$ *is the set of primes* p *for which* $p^nR = p^{n+1}R$ *for some* n, p^e *is the smallest power of* p *for which this holds. Here* $\tau_p(Z(G) \bmod A_{f,Z}(G))$ *stands for the* p-*torsion subgroup of* $Z(G) \bmod A_{f,Z}(G)$ (i.e. *the subgroup* $\{z\in Z(G)\,|\,p^mz \in A_{f,Z}(G)$ *for some* $m \geq 0\}$) *and if* $\sigma(R)$ *is empty, then the right hand side of the above equation is to be interpreted as* $A_{f,Z}(G)$.

<u>Proof</u>. Let $\pi(R)$ denote the set of primes p which are invertible in R

<u>Case I</u>. <u>Characteristic of</u> $R = r \neq 0$.

Since Z/rZ , the ring of integers mod r , can be regarded as a subring of R , $i^{-1}_{Z/rZ}(A_{f,Z/rZ}(G)) \subseteq i_R^{-1}(A_{f,R}(G))$. Let $z \in Z(G)$ be such that $i_R(z) \in A_{f,R}(G)$. If $i_{Z/rZ}(z) \notin A_{f,Z/rZ}(G)$, then $z + A_{f,Z/rZ}(G)$ is a non-zero element of $(Z/rZ)(G)/A_{f,Z/rZ}(G)$. Therefore, we can define a homomorphism

$$\theta : (Z/rZ)(G)/A_{f,Z/rZ}(G) \to T ,$$

where T is the additive group of rationals mod 1, such that

$$\theta(z+A_{f,Z/rZ}(G)) \neq 0 .$$

As the image of θ must be contained in $Z/rZ \subset T$, we have an f_Z-polynomial map $\varphi : G \to Z/rZ$,

$$\varphi(x) = \theta(x+A_{f,Z/rZ}(G)) , \quad x \in G ,$$

such that

$$\varphi^*(z+A_{f,Z}(G)) \neq 0 ,$$

where $\varphi^*: Z(G)/A_{f,Z}(G) \to Z/rZ$ is the homomorphism induced by φ. Composing φ with the imbedding $i : Z/rZ \to R$, we have, by Proposition

1.11, an f_R-polynomial map $\alpha = i\circ\varphi: G \to R$ such that $\alpha^*(z+A_{f,R}(G)) \neq 0$. This is a contradiction, since $z \in A_{f,R}(G)$. Hence $z \in A_{f,Z/rZ}(G)$ and we have

$$i_R^{-1}(A_{f,R}(G)) \subseteq i_{Z/rZ}^{-1}(A_{f,Z/rZ}(G)) .$$

Clearly $i_{Z/nZ}^{-1}(A_{f,Z/nZ}(G)) = A_{f,Z}(G) + nZ(G)$ for every integer $n \geq 1$. Thus the second equality in (i) follows from the observation that if K is a subgroup of an Abelian group H and m, n are coprime integers then $K + mnH = (K+mH) \cap (K+nH)$. Obviously the left hand side is contained in the right hand side. Let $z = k_1 + mh_1 = k_2 + nh_2$, $k_1, k_2 \in K$, $h_1, h_2 \in H$. Then $mh_1 \in K + nH$. Since m, n are coprime, there exist integers a and b such that

$$1 = ma + nb .$$

Therefore, $h_1 = mah_1 + nbh_1 \in K + nH$. Consequently $k_1 + mh_1 \in K + mnH$. Hence

$$(K+mH) \cap (K+nH) \subseteq K + mnH .$$

We now assume that R has characteristic zero.

<u>Case II</u>. <u>$\pi(R)$ = the set of all primes</u>.

In this case $\mathbb{Q}$, the field of rationals, can be regarded as a subring of R. An argument essentially similar to that given in Case I, with $\mathbb{Q}$ in place of both Z/rZ and T, shows that

$$i_R^{-1}(A_{f,R}(G)) = i_{\mathbb{Q}}^{-1}(A_{f,\mathbb{Q}}(G)) .$$

Now

$$i_{\mathbb{Q}}^{-1}(A_{f,\mathbb{Q}}(G)) = \sum_{p\in\pi(R)} \tau_p(Z(G) \bmod A_{f,Z}(G))$$

and we are done.

We next assume that $\pi(R)$ is <u>not</u> the set of all primes.

<u>Case III</u>. $\sigma(R) = \pi(R)$.

Let $z \in i_R^{-1}(A_{f,R}(G))$. We assert that for some integer m, all of whose prime divisors are in $\sigma(R)$, $mz \in A_{f,Z}(G)$. For otherwise, $\bar{z} = z + A_{f,Z}(G)$ is a non-zero element of $Z(G)/A_{f,Z}(G)$ whose order is either infinite or a number at least one of whose prime divisors does not belong to $\sigma(R)$. In either case we can choose a prime $p \notin \sigma(R)$ and a non-zero homomorphism

$$\theta : \langle\bar{z}\rangle \to Z(p^\infty)$$

from the subgroup $\langle\bar{z}\rangle$ generated by $\bar{z}$ into $Z(p^\infty)$ (= the p-subgroup of T). Since $Z(p^\infty)$ is a divisible Abelian group, there exists a homomorphism

$$\varphi : Z(G)/A_{f,Z}(G) \to Z(p^\infty)$$

such that $\varphi(\bar{z}) = \theta(\bar{z})$. Let R_p be the localization of R at the set $\{1,p,p^2,\ldots\}$. Let M be the R-module obtained by factoring R_p with the natural image of R . Then M is an R-module containing $Z(p^\infty)$. Now the map $\alpha : G \to Z(p^\infty)$ given by

$$\alpha(g) = \varphi(g+A_{f,Z}(G)) , g \in G ,$$

is an f_Z-polynomial map. Composing α with the inclusion $Z(p^\infty) \to M$, we obtain (Proposition 1.11) an f_R-polynomial map

$$\alpha : G \to M .$$

Let $\alpha^*: R(G) \to M$ be the linear extension of α to $R(G)$. Then

$$\alpha^*(i_R(z)) = \theta(\bar{z}) \neq 0 .$$

However, $z \in i_R^{-1}(A_{f,R}(G))$. Therefore,

$$\alpha^*(i_R(z)) = 0 .$$

This contradiction proves the assertion. As $\sigma(R) = \pi(R)$, the proof of this case is complete.

<u>Case IV</u>. $\sigma(R) - \pi(R)$ <u>is finite</u>.

We proceed by induction on the order of the set $\sigma(R) - \pi(R)$. When the order is zero, we have the situation of case III. Let $p \in \sigma(R)$, $p \notin \pi(R)$ and let p^e be the smallest power of p for which $p^eR = p^{e+1}R$. Then

(i) $R \cong R/p^eR \oplus R/J$, where $J = \{r \in R | p^e r = 0\}$,

(ii) $\sigma(R) = \sigma(R/J)$

and (iii) p is invertible in R/J .

For (i) consider the homomorphism

$$\varphi : R \to R/p^eR \oplus R/J$$

given by

$$\varphi(r) = (r+p^eR, r+J) .$$

If $r \in \operatorname{Ker} \Phi$, then $r = p^e s$ for some $s \in R$ and $p^e r = 0$. Since $p^e \in p^{e+1}R$, it follows that $r = 0$ and, therefore, Φ is a monomorphism. Let $r,s \in R$. Then we can find $v \in R$ such that $p^e(r-s) = p^{2e}v$.

Clearly $$\varphi(r-p^e v) = (r+p^e R, s+J) .$$
Hence φ is an epimorphism. This establishes (i).

Clearly $\sigma(R)$ is contained in $\sigma(R/J)$. Let $q \in \sigma(R/J)$, $q \neq p$. Suppose $q^n R/J = q^{n+1} R/J$. We can find integers a and b such that
$$p^e a + q^{n+1} b = 1 .$$
Now $q^n = q^{n+1} r + j$ for some $r \in R$ and $j \in J$. Therefore
$$q^n = (p^e a + q^{n+1} b)(q^{n+1} r + j) \in q^{n+1} R ,$$
because $p^e j = 0$.

It only remains to show that p is invertible in R/J. Since $p^e = p^{e+1} t$ for some $t \in R$, $t + J$ is the inverse of $p + J$ in R/J.

In view of (ii) and (iii) we can assume that the Theorem holds for R/J. Let $z \in i_R^{-1}(A_{f,R}(G))$. Then

$$z \in i_{R/J}^{-1}(A_{f,R/J}(G)) =$$

$$\sum_{q \in \sigma(R/J) = \sigma(R)} \{\tau_q(Z(G) \bmod A_{f,Z}(G)) \cap i^{-1}_{Z/q^{e(q)}Z}(A_{f,Z/q^{e(q)}Z}(G)\}$$

where $e(q)$ is the least integer for which $q^{e(q)} R/J = q^{e(q)+1} R/J$. It is easy to see that for $q \neq p$, $e(q)$ = the least integer for which
$$q^{e(q)} R = q^{e(q)+1} R .$$
Hence
$$z \in \sum_{\substack{q \in \sigma(R) \\ q \neq p}} \{\tau_q(Z(G) \bmod A_{f,Z}(G)) \cap i^{-1}_{Z/q^{e(q)}Z}(A_{f,Z/q^{e(q)}Z}(G)\} + \tau_p(Z(G) \bmod A_{f,Z}(G)) .$$
Since $z \in i^{-1}_{R/p^e R}(A_{f,R/p^e R}(G)) = i^{-1}_{Z/p^e Z}(A_{f,Z/p^e Z}(G))$ (Case I) and
$$\tau_q(Z(G) \bmod A_{f,Z}(G)) \subseteq i^{-1}_{Z/p^e Z}(A_{f,Z/p^e Z}(G)) ,$$
it follows that

$$z \in \sum_{q \in \sigma(R)} \tau_q(Z(G) \bmod A_{f,Z}(G)) \cap i^{-1}_{Z/q^{e(q)}Z}(A_{f,Z/q^{e(q)}Z}(G)) .$$

Conversely, if $z \in \sum_{q \in \sigma(R)} \{\tau_q(Z(G) \bmod A_{f,Z}(G)) \cap i^{-1}_{Z/q^{e(q)}Z}(A_{f,Z/q^{e(q)}Z}(G)\}$ then, by induction, $z \in i^{-1}_{R/J}(A_{f,R/J}(G))$ and also $z \in i^{-1}_{R/p^e R}(A_{f,R/p^e R}(G)) =$

$i^{-1}_{Z/p^eZ}(A_{f,Z/p^eZ}(G))$. Hence $i_R(z) = x + y$, where $x \in A_{f,R}(G)$ and $y \in R(G)$ has all its coefficients in J so that $p^e y = 0$. Also $i_R(z) = u + p^e v$, $u \in A_{f,R}(G)$, $v \in R(G)$. Since $p^e R = p^{e+1} R$, it follows that $i_R(z) \in A_{f,R}(G)$.

Case V. $\sigma(R)$ is arbitrary.

Let $z \in Z(G)$ be such that $i_R(z) \in A_{f,R}(G)$. By considering an expression of $i_R(z)$ as an element of $A_{f,R}(G)$, it follows that there is a finitely generated subring S of R such that

$$i_S(z) \in A_{f,S}(G) .$$

Suppose we have proved the Theorem for finitely generated rings. Then

$$z \in \sum_{p\in\sigma(S)} \{\tau_p(Z(G) \bmod A_{f,Z}(G)) \cap i^{-1}_{Z/p^{e'}Z}(A_{f,Z/p^{e'}Z}(G))\}$$

where $p^{e'}$ is the smallest power of p for which $p^{e'}S = p^{e'+1}S$. Now $p \in \sigma(S)$ implies that $p \in \sigma(R)$ and if p^e is the smallest power of p for which $p^e R = p^{e+1} R$, then $e \leq e'$. Hence

$$z \in \sum_{p\in\sigma(R)} \{\tau_p(Z(G) \bmod A_{f,R}(G)) \cap i^{-1}_{Z/p^eZ}(A_{f,Z/p^eZ}(G))\} .$$

Consequently

$$i_R^{-1}(A_{f,R}(G)) \subseteq \sum_{p\in\sigma(R)} \{\tau_p(Z(G) \bmod A_{f,Z}(G)) \cap i^{-1}_{Z/p^eZ}(A_{f,Z/p^eZ}(G))\} .$$

For the reverse inclusion, let

$$z \in \tau_p(Z(G) \bmod A_{f,Z}(G)) \cap i^{-1}_{Z/p^eZ}(A_{f,Z/p^eZ}(G))$$

for some $p \in \sigma(R)$. Then $i_R(z) = u + p^e v$, $u \in A_{f,R}(G)$, $v \in R(G)$ and there exists an integer m such that

$$p^m i_R(z) \in A_{f,R}(G) .$$

Since $p^e R = p^{e+1} R$, it follows that

$$i_R(z) \in A_{f,R}(G) .$$

Finally, let R be a finitely generated commutative ring with identity. We assert that $\sigma(R) - \pi(R)$ must be finite. For, suppose we can find an infinity of primes $p_1, p_2, \ldots, p_i, \ldots$ in $\sigma(R) - \pi(R)$. Let $p_i^{e(p_i)}$ be the smallest power of p_i such that $p_i^{e(p_i)} R = p_i^{e(p_i)+1} R$. Let $t_m = \prod_{1\leq i\leq m} p_i^{e(p_i)}$ and $J_m = \{r \in R / t_m r = 0\}$. Then $J_1 \subseteq J_2 \subseteq \cdots \subseteq J_m \subseteq \cdots$ For each m there is a $v_m \in R$ such that $p_m^{e(p_m)} = p_m^{e(p_m)+1} v_m$. The element $1 - p_m v_m \in J_m$ and $\notin J_{m-1}$ because p_m is not invertible in R . Hence the above series is a strictly increasing series of ideals in R . As R is Noetherian (being finitely generated), this is not possible. Hence $\sigma(R) - \pi(R)$ must be finite and the proof of Theorem 1.12 is complete.

CHAPTER II

DIMENSION SUBGROUPS

Every two sided ideal of a group ring R(G) determines a normal subgroup of the group G. Of special interest are the (Lie) dimension subgroups determined by the (Lie) associative powers of the augmentation ideal $\Delta_R(G)$. This Chapter reduces the study of these subgroups over arbitrary coefficient rings to the integral and modular coefficients.

1. SUBGROUPS DETERMINED BY POLYNOMIAL IDEALS

In this Chapter R denotes a commutative ring with identity, G denotes a multiplicatively written group and R(G) the group ring of G over R.

If N is a normal subgroup of G, then $\Delta_R(G,N)$ is a two-sided ideal of R(G). This gives a map

$$\varphi : \Sigma \to \mathfrak{I}, \quad \varphi(N) = \Delta_R(G,N)$$

where Σ is the set of normal subgroups of G and $\mathfrak{I}$ is the set of two-sided ideals of R(G). On the other hand, if I is a two-sided ideal of R(G), then $G \cap (1+I)$ is easily seen to be a normal subgroup of G. This leads to a map

$$\psi : \mathfrak{I} \to \Sigma, \quad \psi(I) = G \cap (1+I).$$

Since $\Delta_R(G,N)$, by definition, is the kernel of the ring homomorphism $R(G) \to R(G/N)$ induced by the natural projection $G \to G/N$, it is clear that

$$g - 1 \in \Delta_R(G,N) \Rightarrow g \in N.$$

Thus

$$\psi \circ \varphi = \text{identity}.$$

This, however, is not true for the composition $\varphi \circ \psi$. For, consider R(G) itself. Evidently $\psi(R(G)) = G$ and $\varphi(G) = \Delta_R(G,G) = \Delta_R(G)$.

The above correspondence between normal subgroups of G and two-sided ideals of group rings R(G) of G is a rich source of results both for the group and the group ring and has been investigated by several authors over the last four decades.

We are primarily interested in the normal subgroups $\psi(\Delta_R^i(G))$ and $\psi(\Delta_R^{(i)}(G))$, $i \geq 1$.

1.1 Definition. The normal subgroup $\psi(\Delta_R^i(G))$ is called the *i-th dimension subgroup of* G *over* R and is denoted by $D_{i,R}(G)$. The normal subgroup $\psi(\Delta_R^{(i)}(G))$ is called the *i-th Lie dimension subgroup of* G *over* R and is denoted by $D_{(i),R}(G)$.

As has been shown in Chapter I, the ideals $\Delta_R^i(G)$ and $\Delta_R^{(i)}(G)$ are polynomial ideals for all $i \geq 1$. For normal subgroups determined by plynomial ideals, we have

1.2 Proposition. *If* $A_{f,R}(G)$ *is a polynomial ideal of* $R(G)$, $\theta : G \to H$ *is a group homomorphism and* $\varphi : R \to S$ *is a ring homomorphism, then*

$$\theta(\psi(A_{f,R}(G)) \subseteq \psi(A_{f,S}(H)) .$$

Proof. Extend $\theta : G \to H$ and $\varphi : R \to S$ to a ring homomorphism $\alpha^* : R(G) \to S(H)$, $\alpha^*(\sum_{g \in G, r_g \in R} r_g g) = \Sigma\varphi(r_g)\theta(g)$. Then $\alpha^*(A_{f,R}(G)) \subseteq A_{f,S}(H)$ [Chapter I, Proposition 1.4]. Thus if $g \in G \cap (1+A_{f,R}(H))$, then $\theta(g) \in H \cap (1+A_{f,S}(H))$.

1.3 Corollary. *If* I *is a polynomial ideal of* $R(G)$, *then* $\psi(I)$ *is a fully invariant subgroup of* G . [A subgroup H of a group G is said to be fully invariant if for every endomorphism $\theta : G \to G$, $\theta(H) \subseteq H$.]

In order to give the reader a flavour of how the group and ring properties are reflected in the subgroups $\psi(I)$, we calculate $D_{2,R}(T)$, where T is the group $\mathbb{Q}/\mathbb{Z}$, $\mathbb{Q}$ = the additive group of rationals. It may be recalled that $T = \Sigma Z(p^\infty)$, direct sum over all primes p . For our purposes, we must write T *multiplicatively*.

1.4 Example [59]. *For every commutative ring* R *with identity*, $D_{2,R}(T) = \sum_{p \in \sigma(R)} Z(p^\infty)$ *where* $\sigma(R) = \{p | p$ *is a prime and* $p^n R = p^{n+1} R$ *for some* $n \geq 0\}$ [If $\sigma(R)$ is empty, the right hand side is to be interpreted as the identity subgroup.]

Proof. Let $p \in \sigma(R)$, $t \in Z(p^\infty)$. Then there exists $n \geq 0$ and $x \in Z(p^\infty)$ such that $p^n R = p^{n+1} R$ and $t = x^{p^n}$ (recall that $Z(p^\infty)$ is a divisible Abelian group which we are writing multiplicatively.) Now

$$t - 1 = x^{p^n} - 1$$
$$= p^n(x-1) + \binom{p^n}{2}(x-1)^2 + \ldots + (x-1)^{p^n}$$
$$\equiv p^n(x-1) \mod \Delta_R^2(Z(p^\infty)) \quad , \text{ since}$$
$$(x-1)^r \in \Delta_R^2(Z(p^\infty)) \quad \text{for} \quad r \geq 2 .$$

Let order of x be p^m. Then the equation

$$0 = x^{p^m} - 1 = p^m(x-1) + \binom{p^m}{2}(x-1)^2 + \ldots + (x-1)^{p^m}$$

shows that $p^m(x-1) \in \Delta_R^2(Z(p^\infty))$. Since $p^nR = p^{n+1}R$, there exists $r \in R$ such that $p^n = p^{n+m}r$. Hence $p^n(x-1) = rp^n \cdot p^m(x-1) \in \Delta_R^2(Z(p^\infty))$. Consequently $t - 1 \in \Delta_R^2(Z(p^\infty))$. Hence

$$\sum_{p \in \sigma(R)} Z(p^\infty) \subseteq D_{2,R}(T) .$$

Next let $t \in D_{2,R}(T)$. Then for any prime p, t_p, the p-primary component of t, is in $D_{2,R}(Z(p^\infty))$. This follows (Proposition 1.2) from the projection of T on its direct summand $Z(p^\infty)$. Let H be the subgroup generated by the elements of $Z(p^\infty)$ which appear in an expression of the type $\sum_{r \in R, x, y \in Z(p^\infty)} r(x-1)(y-1)$ for $t_p - 1$ as an element of $\Delta_R^2(Z(p^\infty))$.
Then $(Z(p^\infty)$ being locally cyclic) H is a cyclic group of order p^r, say, and $t_p \in D_{2,R}(H)$. Let $p \notin \sigma(R)$. Then the ring $S(n) = R/p^nR$, $n \geq 1$, has characteristic p^n. As $D_{2,R}(H) \subseteq D_{2,S(n)}(H)$, we have $t_p \in D_{2,S(n)}(H)$ for all $n \geq 1$. Let H be generated by h and $t_p = h^{p^s}$, $r \geq s \geq 0$. Then $t_p - 1 \in \Delta_{S(r)}^2(H)$ implies that

$$h^{p^s} - 1 = p^s(h-1) + \binom{p^s}{2}(h-1)^2 + \ldots + (h-1)^{p^s} \in \Delta_{S(r)}^2(H) .$$

Therefore, $p^s(h-1) \in \Delta_{S(r)}^2(H)$. Clearly $\Delta_{S(r)}^2(H)$ is generated as an ideal by $(h-1)^2$. Hence there exists $u \in S(r)(H)$ such that

$$p^s(h-1) = (h-1)^2 u .$$

Let $\alpha = p^s - (h-1)u$. Then $h\alpha = \alpha$. Therefore, $\alpha = v(1+h+h^2+\ldots+h^{p^r-1})$ for some $v \in S(r)$ (compare coefficients of h^i, $0 \leq i < p^r$, on both sides of $h\alpha = \alpha$). Applying the unit augmentation $\varepsilon : S(r)(H) \to S(r)$ to both sides of the equation

$$p^s - (h-1)u = v(1+h+h^2+\ldots+h^{p^r-1})$$

we get

$$p^s = 0 . \quad (\text{in } S(r)) .$$

Hence $s \geq r$ and $t_p = 1$. Thus if $t \in D_{2,R}(T)$, then for every $p \notin \sigma(R)$, $t_p = 1$, i.e.

$$t \in \sum_{p \in \sigma(R)} Z(p^{\infty}) .$$

This completes the proof.

As another example we compute the second integral and modular dimension subgroups of an arbitrary group.

1.5 <u>Example</u>. <u>For every integer</u> $m \geq 0, D_{2,Z/mZ}(G) = \gamma_2(G)G^m$, <u>where</u> G^m <u>is the subgroup of</u> G <u>generated by</u> x^m $(x \in G)$.

<u>Proof</u>. Consider the map $u : G \to \Delta_Z(G)/\Delta_Z^2(G) + m\Delta_Z(G) \cong \Delta_{Z/mZ}(G)/\Delta_{Z/mZ}^2(G)$ given by $u(x) = x - 1 + \Delta_Z^2(G) + m\Delta_Z(G)$. Clearly u is a homomorphism and $\gamma_2(G)G^m \subseteq \operatorname{Ker} u = D_{2,Z/mZ}(G)$. Thus we have a homomorphism $u^*: G/\gamma_2(G)G^m \to \Delta_Z(G)/\Delta_Z^2(G) + m\Delta_Z(G)$. Now the ideal $\Delta_Z(G)$ is a free Abelian group with the elements $g-1$ $(1 \neq g \in G)$ as a basis. We define $v : \Delta_Z(G) \to G/\gamma_2(G)G^m$ by $v(g-1) = g\,\gamma_2(G)G^m$. Then $\Delta_Z^2(G) + m\,\Delta_Z(G) \subseteq \operatorname{Ker} v$ and we have a homomorphism $v^*: \Delta_Z(G)/\Delta_Z^2(G) + m\,\Delta_Z(G) \to G/\gamma_2(G)G^m$. By definitions $u^* o v^* =$ identity and $v^* o u^* =$ identity. Hence $\operatorname{Ker} u^* = 1$ and so $\operatorname{Ker} u = D_{2,Z/mZ}(G) = \gamma_2(G)G^m$.

We next prove two reduction Theorems which show that for the study of the subgroups $D_{i,R}(G)$ and $D_{(i),R}(G)$ it is enough to confine one's attention to the cases where R is Z, the ring of rational integers or Z/p^eZ, p prime.

2. <u>DIMENSION SUBGROUPS OVER ARBITRARY RINGS OF COEFFICIENTS</u>

2.1 <u>Theorem</u> ([59], [80]; see also [57]). (i) <u>If characteristic of</u> R <u>is zero, then</u>

$$D_{n,R}(G) = \prod_{p \in \sigma(R)} \{\tau_p(G \bmod D_{n,Z}(G)) \cap D_{n,Z/p^eZ}(G)\}$$

<u>where</u> $\sigma(R) = \{p \mid p$ <u>is a prime and</u> $p^nR = p^{n+1}R$ <u>for some</u> $n \geq 0\}$ <u>and for</u> $p \in \sigma(R)$, p^e <u>is the smallest power of</u> p <u>for which</u> $p^eR = p^{e+1}R$. [<u>If</u> $\sigma(R)$ <u>is empty, then the right hand side is to be interpreted as</u> $D_{n,Z}(G)$.]

(ii) <u>If characteristic of</u> R <u>is</u> $r > 0$, <u>then, for all</u> $n \geq 1$, $D_{n,R}(G) = D_{n,Z/rZ}(G) = \bigcap_i D_{n,Z/p_i^{e_i}Z}(G)$ <u>where</u> $r = \prod p_i^{e_i}$ <u>is the prime factorization of</u> r.

<u>Proof</u>. Suppose characteristic of R is zero. Let $g \in D_{n,R}(G)$. Then

$g - 1_R \in \Delta_R^n(G)$, where 1_R is the identity of R. Let $f(x_1,x_2,\dots,x_n) = (x_1-1)(x_2-1)\dots(x_n-1)$. Then $A_{f,R}(G) = \Delta_R^n(G)$. Therefore, by [Chapter I, Theorem 1.12], we have

$$g - 1 = \sum_{p\in\sigma(R)} z_p$$

where $z_p \in Z(G)$ is such that, for some $m = m(p) \geq 0$, $p^m z_p \in \Delta_Z^n(G)$ and $i_{Z/p^eZ}(z_p) \in \Delta^n_{Z/p^eZ}(G)$. Let $r = \Pi p^{m(p)}$. Then r is a σ-number (i.e. all its prime divisors belong to $\sigma(R)$) and $r(g-1) \in \Delta_Z^n(G)$. We assert that the order of $g \bmod D_{n,Z}(G)$ is a σ-number. It is easy to see that if $u = p^\alpha\cdot\beta$, $v = p^{\alpha'}\cdot\beta'$, $u \geq v \geq 1$, $p\nmid\beta,\beta'$, then the exact power of p which divides the binomial coefficient $\binom{u}{v}$ is $p^{\alpha-\alpha'}$. Let, for every prime p, $p^{a(p)}$ be the highest power of p which divides any of the integers $1,2,\dots,n-1$. Choose an integer s such that $(s+1)m(p) \geq a(p)$ for all primes p with $m(p) > 0$. Then clearly $p^{m(p)}$ divides $\binom{r^s}{i}$ for every $i = 1,2,\dots,n-1$. Hence r divides the integers $\binom{r^s}{i}$, $i = 1,2,\dots,n-1$. Consequently

$$g^{r^s} - 1 = \sum_{i=1}^{r^s} \binom{r^s}{i}(g-1)^i$$

$$\equiv 0 \bmod \Delta_Z^n(G)$$

i.e. $g^{r^s} \in D_{n,Z}(G)$ and so the order of $g \bmod D_{n,Z}(G)$ is a σ-number. Let $p_1,p_2,\dots,p_k$ be all the primes $p \in \sigma(R)$ with $m(p) \neq 0$. Let $q_i = r^s/p_i^{sm(p_i)}$, $i = 1,2,\dots,k$. Then the greatest common divisor of the integers $q_1,q_2,\dots,q_k$ is 1. Therefore, there exist integers u_i, $1 \leq i \leq k$, such that

$$1 = \sum_{i=1}^{k} q_i u_i .$$

Let $g_i = g^{q_i u_i}$. Then $g_i^{p_i^{sm(p_i)}} = g^{r^s u_i} \in D_{n,Z}(G)$ i.e. $g_i \in \tau_{p_i}(G \bmod D_{n,Z}(G))$. Since $i_{Z/p^eZ}(\tau_q(Z(G) \bmod \Delta_Z^n(G)) \subseteq \Delta^n_{Z/p^eZ}(G)$ for all $q \neq p$, it is clear that $g \in D_{n,Z/p^eZ}(G)$ for all $p \in \sigma(R)$. Hence $g_i = g^{q_i u_i} \in D_{n,Z/p^eZ}(G)$. Consequently

$$g = \prod_{i=1}^{k} g_i \in \prod_{p\in\sigma(R)} \{\tau_p(G \bmod D_{n,Z}(G)) \cap D_{n,Z/p^eZ}(G)\} .$$

Conversely, let $g \in \tau_p(G \bmod D_{n,Z}(G)) \cap D_{n,Z/p^eZ}(G)$, $p \in \sigma(R)$. Then for some $u \geq 0$, $g^{p^u} \in D_{n,Z}(G)$. If H is the subgroup generated by g , then the equation

$$(2.2) \qquad g^{p^u} - 1 = p^u(g-1) + \binom{p^u}{2}(g-1)^2 + \ldots + (g-1)^{p^u}$$

shows that

$$p^u(g-1) \in \Delta_Z^2(H) + \Delta_Z^n(G) .$$

Therefore,

$$p^{(n-1)u}(g-1) \in \Delta_Z^n(G) .$$

Thus

$$g - 1 \in \tau_p(Z(G) \bmod \Delta_Z^n(G)) \cap i^{-1}_{Z/p^eZ}(\Delta^n_{Z/p^eZ}(G)) .$$

Hence, by [Chapter I, Theorem 1.12], $g - 1_R \in \Delta_R^n(G)$, i.e. $g \in D_{n,R}(G)$. This completes the proof of case (i).

Case (ii) follows immediately from [Chapter I, Theorem 1.12 (i)].

3. LIE DIMENSION SUBGROUPS OVER ARBITRARY RINGS OF COEFFICIENTS

3.1 Theorem [59]. (i) If characteristic of R is zero, then

$$D_{(n),R}(G) = \prod_{p \in \sigma(R)} \gamma_2(G) \cap \{\tau_p(G \bmod D_{(n),R}(G)) \cap D_{(n),Z/p^eZ}(G)$$

for $n \geq 2$, where $\sigma(R)$ and p^e are as defined in Theorem 2.1. [If $\sigma(R)$ is empty, the right hand side is to be interpreted as $D_{(n),Z}(G)$. Here $\gamma_2(G)$ = the derived group of G .]

(ii) If characteristic of R is $r > 0$, then, for all $n \geq 1$,

$$D_{(n),R}(G) = D_{(n),Z/rZ}(G) = \bigcap_i D_{(n),Z/p_i^{e_i}Z}(G) ,$$

where $r = \prod p_i^{e_i}$ is the prime factorization of r .

Proof. [The proof of this Theorem is similar to that of Theorem 2.1. We have, therefore, somewhat condensed the arguments.]

Suppose characteristic of R is zero. Since $\Delta_R^{(2)}(G) = \Delta_R(G,\gamma_2(G))$ it is clear that $D_{(n),R}(G) \subseteq \gamma_2(G)$ for all $n \geq 2$. Let $g \in D_{(n),R}(G)$, $n \geq 2$. As $\Delta_R^{(n)}(G)$ is a polynomial ideal [Chapter I, Theorem 1.12] shows that for some σ-number r , $r(g-1) \in \Delta_Z^{(n)}(G)$. By [Chapter I, Theorem 1.8], $(g-1)^m \in \Delta_Z^{(m+1)}(G)$ for all m . Hence, choosing s sufficiently large, we can conclude (by (2.2)) that $g^{r^s} - 1 \in \Delta_Z^{(n)}(G)$

which yields that the order of $g \bmod D_{(n),Z}(G)$ is a σ-number. Hence $g = g_1 g_2 \cdots g_k$ where each g_i is a power of g and is a p-element mod $D_{(n),Z}(G)$ for some $p \in \sigma(R)$. Thus

$$g \in \prod_{p \in \sigma(R)} \gamma_2(G) \cap \{\tau_p(G \bmod D_{(n),Z}(G)) \cap D_{(n),Z/p^eZ}(G)\} .$$

Conversely, let $g \in \gamma_2(G) \cap \{\tau_p(G \bmod D_{(n),Z}(G) \cap D_{(n),Z/p^eZ}(G)\}$, $p \in \sigma(R)$. Then $g^{p^r} - 1 \in \Delta_Z^{(n)}(G)$ for some r. As $(g-1)^m \in \Delta_Z^{(m+1)}(G)$ for all m, we can find (by using (2.2)) an s such that $p^s(g-1) \in \Delta_Z^{(n)}(G)$. Thus $g - 1 \in \tau_p(Z(G) \bmod \Delta_Z^{(n)}(G)) \cap i^{-1}_{Z/p^eZ}(\Delta^{(n)}_{Z/p^eZ}(G))$. Hence by [Chapter I, Theorem 1.12], $g - 1_R \in \Delta_R^{(n)}(G)$, i.e. $g \in D_{(n),R}(G)$. This completes the proof of case (i). Case (ii) follows from [Chapter I, Theorem 1.12 (ii)].

CHAPTER III

GROUP RINGS OF NILPOTENT GROUPS

In this Chapter we study a correspondence between N-series and their associated canonical filtrations of augmentation ideals. The main part of the Chapter is devoted to an exposition of a construction in the group rings of nilpotent groups due to B. Hartley [27]. Our main deductions from Hartley's theory are Theorems 1.6 and 1.7.

1. N-SERIES AND FILTRATIONS OF THE AUGMENTATION IDEAL

1.1 Definition. A series $G = H_1 \supseteq H_2 \supseteq \dots \supseteq H_i \supseteq \dots$ of subgroups of a group G is called an N-series if $(H_i, H_j) \subseteq H_{i+j}$ for all $i, j \geq 1$, where (M,N) denotes the subgroup generated by all commutators $(m,n) = m^{-1}n^{-1}mn$, $m \in M$, $n \in N$. An N-series $\{H_i\}_{i \geq 1}$ is called a restricted N-series relative to a prime p if $x \in H_i$ implies that $x^p \in H_{ip}$ for all $i \geq 1$, i.e. $H_i^p \subseteq H_{ip}$ for all $i \geq 1$.

The most familiar example of an N-series in a group is its lower central series $\{\gamma_i(G)\}$. The lower central series has the property that if $\{H_i\}$ is any N-series in G, then $\gamma_i(G) \subseteq H_i$ for all $i \geq 1$.

Let G be a group and R a ring with identity. Then, just as in the commutative case, we can define the group ring $R(G)$ of G over R and its augmentation ideal $\Delta_R(G)$ [all R-modules, when R is non-commutative, will be left R-modules]. By a filtration of the augmentation ideal $\Delta_R(G)$ we mean a decreasing series

$$\Delta_R(G) = A_1 \supseteq A_2 \supseteq \dots \supseteq A_n \supseteq \dots$$

of two-sided ideals A_n of $R(G)$.

A natural way in which N-series arise is from filtrations of the augmentation ideals.

1.2 Proposition. *Let G be a group and R a ring with identity. Let $\Delta_R(G) = A_1 \supseteq A_2 \supseteq \dots \supseteq A_n \supseteq \dots$ be a filtration of $\Delta_R(G)$ such that $A_iA_j \subseteq A_{i+j}$ for all $i, j \geq 1$. If $H_i = G \cap (1+A_i)$, then $\{H_i\}_{i \geq 1}$ is an N-series. Further, if the characteristic of R is a prime p, then $\{H_i\}_{i \geq 1}$ is a restricted N-series relative to p.*

Proof. Let $x \in H_i$, $y \in H_j$. Then, by definition, $x - 1 \in A_i$, $y - 1 \in A_j$. Therefore, $(x-1)(y-1)$ and $(y-1)(x-1)$ both belong to A_{i+j}. Hence

$$(x,y) - 1 = x^{-1}y^{-1}\{(x-1)(y-1) - (y-1)(x-1)\} \in A_{i+j} .$$

This proves that $\{H_i\}$ is an N-series. If the characteristic of R is p, then we have the following identity in $R(G)$:

$$(x-1)^p = x^p - 1 .$$

Thus if $x \in H_i$, then $x^p - 1 \in A_{ip}$. Hence $H_i^p \subseteq H_{ip}$ and $\{H_i\}$ is a restricted N-series relative to p.

A problem, extensively investigated by Lazard [38], is the characterization of those N-series of a group G which are determined by filtrations of $R(G)$ as in Proposition 1.2. In fact, Lazard proved that every restricted N-series is obtained in this way (see Theorem 1.7).

The filtration of $\Delta_R(G)$ given by the (associative) powers $\Delta_R^i(G)$ of its augmentation ideal obviously has the property that

$$\Delta_R^i(G)\cdot\Delta_R^j(G) \subseteq \Delta_R^{i+j}(G) \quad \text{for all} \quad i,j \geq 1 .$$

Consequently, we have the following

1.3 <u>Corollary</u>. <u>If</u> G <u>is a group and</u> R <u>is a ring with identity, then</u>

$$G = D_{1,R}(G) \supseteq D_{2,R}(G) \supseteq \cdots \supseteq D_{n,R}(G) \supseteq \cdots$$

<u>is an N-series</u>. <u>Further, if the characteristic of</u> R <u>is a prime</u> p, <u>then this series is a restricted N-series relative to</u> p. <u>In particular</u>, $\gamma_n(G) \subseteq D_{n,R}(G)$ <u>for all</u> $n \geq 1$. [Here $D_{n,R}(G) = G \cap (1+\Delta_R^n(G))$].

Every N-series $\{H_i\}$ of G induces a <u>weight function</u> w on G defined as follows:

$$(1.4) \qquad w(x) = \begin{cases} k & \text{if} \quad x \in H_k \setminus H_{k+1} \\ \infty & \text{if} \quad x \in \bigcap_i H_i \end{cases} .$$

Since $(H_i,H_j) \subseteq H_{i+j}$, w satisfies:

$$w((x,y)) \geq w(x) + w(y) \quad \text{for all} \quad x,y \in G .$$

1.5 <u>The canonical filtration induced by an N-series</u>

If R is a ring with identity and G is a group, then every N-series $\{H_i\}$ of G induces a filtration of $\Delta_R(G)$ as follows:

For $n \geq 1$, define A_n to be the R-submodule of $R(G)$ spanned by all products $(g_1-1)(g_2-1)\dots(g_s-1)$ with $\sum_{i=1}^{s} w(g_i) \geq n$. Then $A_1 = \Delta_R(G)$ and $A_iA_j \subseteq A_{i+j}$ for all $i,j \geq 1$. Thus each A_n, $n \geq 1$, is a two-sided ideal of $R(G)$ and we have a filtration $\Delta_R(G) = A_1 \supseteq A_2 \supseteq \dots \supseteq A_n \supseteq \dots$. We call this filtration $\{A_n\}$ <u>the canonical filtration of</u> $R(G)$ <u>induced by the N-series</u> $\{H_i\}$.

The canonical filtration of $R(G)$ induced by the lower central series $\{\gamma_i(G)\}$ is the filtration given by the powers of the augmentation ideal; i.e. $A_n = \Delta_R^n(G)$ for all $n \geq 1$.

Let $\{H_i\}$ be an N-series of a group G and let $\{A_n\}$ be the canonical filtration of $\Delta_R(G)$ induced by $\{H_i\}$. We are interested in the investigation of conditions under which the N-series determined by $\{A_n\}$ is $\{H_i\}$ itself, i.e. $H_i = G \cap (1+A_i)$ for all $i \geq 1$. The so-called <u>dimension subgroup problem</u> corresponds to the case when $\{H_i\}$ is the lower central series of G and R is the ring Z of integers [For the present status of this problem see Chapter V.]

We prove here two results which will be used to determine the dimension subgroups over fields.

1.6 <u>Theorem</u>. Let $G = H_1 \supseteq H_2 \supseteq \dots \supseteq H_c \supseteq H_{c+1} = 1$ <u>be a finite N-series with</u> H_i/H_{i+1} <u>torsion-free for</u> $i = 1,2,\dots,c$. <u>Let</u> R <u>be a ring with identity having zero characteristic</u>. <u>Then</u>

$$H_i = G \cap (1+A_i) \quad \text{for all} \quad i \geq 1 ,$$

<u>where</u> $\{A_i\}$ <u>is the canonical filtration of</u> $R(G)$ <u>induced by</u> $\{H_i\}$.

1.7 <u>Theorem</u>. <u>Let</u> $G = H_1 \supseteq H_2 \supseteq \dots \supseteq H_c \supseteq H_{c+1} = 1$ <u>be a finite restricted N-series relative to a prime</u> p. <u>Let</u> $\{A_i\}$ <u>be the canonical filtration of</u> $\Delta_R(G)$ <u>where</u> R <u>is a ring (with identity) of characteristic</u> p. <u>Then</u>

$$H_i = G \cap (1+A_i) \quad \text{for all} \quad i \geq 1 .$$

Theorem 1.6 is implicit in the work of Hall [26] and Jennings [36] while Theorem 1.7 is implicit in the work of Lazard [38]. We propose to follow the fundamental paper of Hartley [27] in developing the machinery which handles both cases simultaneously.

2. A CONSTRUCTION IN GROUP RINGS OF NILPOTENT GROUPS

Let

$$G = G_1 \supseteq G_2 \supseteq \dots \supseteq G_c \supseteq G_{c+1} = 1 \tag{2.1}$$

be a finite N-series of a group G such that each of the Abelian groups G_i/G_{i+1} is a direct sum of (possibly infinitely many) cyclic groups which are either infinite or of order p^k, where p is a fixed prime and k is bounded by some integer K depending only on G. Thus there exist ordinals

$$0 = \lambda_{c+1} \leq \lambda_c \leq \ldots \leq \lambda_1 = \lambda$$

and elements $(x_\alpha)_{\alpha<\lambda}$ in G such that G_i/G_{i+1} is the direct sum of the non-trivial cyclic groups generated by the cosets $x_\alpha G_{i+1}$, $\lambda_{i+1} \leq \alpha < \lambda_i$. By our assumptions, if $\lambda_{i+1} \leq \alpha < \lambda_i$, then $x_\alpha G_{i+1}$ has either infinite order or order p^{n_α} with $0 < n_\alpha \leq K$. For $\alpha < \lambda$, we define $\mu(\alpha)$ by

(2.2) $$\mu(\alpha) = w(x_\alpha)$$

where w is the weight function (1.4) defined by the given N-series on G. Then each element $x \in G$ may be written uniquely in the form

(2.3) $$x = x_{\alpha_1}^{r_{\alpha_1}} \cdots x_{\alpha_k}^{r_{\alpha_k}}$$

where $k \geq 0$, $\alpha_1 < \alpha_2 < \ldots < \alpha_k < \lambda$, $0 \neq r_{\alpha_i} \in Z$ and $0 < r_{\alpha_i} < p^{n_{\alpha_i}}$ if $x_{\alpha_i} G_{\mu(\alpha_i)+1}$ has finite order $p^{n_{\alpha_i}}$. We call $\Phi = (x_\alpha)_{\alpha<\lambda}$ a <u>canonical basis</u> of G.

Consider the set $\underset{\sim}{S}$ of vectors $\underset{\sim}{r} = (r_\alpha)_{\alpha<\lambda}$, $r_\alpha \in Z$, such that almost all r_α are zero and $0 \leq r_\alpha < p^{n_\alpha}$ if $x_\alpha G_{\mu(\alpha)+1}$ has order p^{n_α}.

Let M be a fixed natural number and R a ring with identity. With each $\underset{\sim}{r} \in \underset{\sim}{S}$ and non-zero $\theta \in R$, we associate an element $u(\underset{\sim}{r},\theta) \in R(G)$, namely

(2.4) $$u(\underset{\sim}{r},\theta) = \theta \prod_{\alpha<\lambda} v_\alpha$$

where, if $u_\alpha = 1 - x_\alpha$, then $v_\alpha = v_\alpha(\underset{\sim}{r})$ is given by

(2.5) $$v_\alpha = \begin{cases} u_\alpha^{r_\alpha} & \text{if } r_\alpha \geq 0 \\ u_\alpha x_\alpha^{M r_\alpha} & \text{if } r_\alpha < 0 \end{cases}$$

and the product in (2.4) is taken in natural order. We write

$$u(\underset{\sim}{r}) = u(\underset{\sim}{r},1).$$

We assert that the elements $u(\underset{\sim}{r})$ span $R(G)$ over R (as a left R-module). For this we need the following result which is essentially Lemma 7.2 of [26].

2.6 Lemma. Let $\{x\}$ be a cyclic group. R a ring with identity and u the element $1 - x$ of the group ring $R\{x\}$. Then

(i) for all $n \geq 0$ the subsets $\{1,u,u^2,\ldots,u^n\}$ and $\{1,x,x^2,\ldots,x^n\}$ span the same R-submodule of $R\{x\}$;

(ii) if x has infinite order and M is a fixed integer ≥ 0, then the elements u^r, $r \geq 0$, and $u^M x^{-s}$, $s > 0$, span $R\{x\}$ over R [Hartley [28] has found a basis of the group ring of an infinite cyclic group which is adapted to the powers of the augmentation ideal and does not involve the number M.]

Proof. For every $r \geq 0$ we have

$$x^r = \sum_{i=0}^{r} \binom{r}{i} (-u)^i .$$

Hence the subset $\{1,x,x^2,\ldots,x^n\}$ is contained in the R-submodule spanned by $\{1,u,u^2,\ldots,u^n\}$ for all $n \geq 0$. The reverse inclusion being immediate, (i) is proved.

To prove (ii) we proceed by induction on M. Let U_M denote the R-submodule spanned by $\{u^r | r \geq 0\} \cup \{u^M x^{-s} | s > 0\}$. We have to show that $U_M = R\{x\}$. That $U_0 = R\{x\}$ follows from (i). Assume that $M > 0$ and $U_{M-1} = R\{x\}$. To prove that $U_M = R\{x\}$ it clearly suffices to show that the elements $u^{M-1}x^{-s}$ lie in U_M for $s > 0$.

Now

$$x^{-s} = x^{-(s-1)} + (1-x)x^{-s}$$

whence

$$u^{M-1}x^{-s} = u^{M-1}x^{-(s-1)} + u^M x^{-s}.$$

Thus induction on $s > 0$ shows that $u^{M-1}x^{-s} \in U_M$ for all $s > 0$.

2.7 Lemma. The elements $u(\underset{\sim}{r})$, $\underset{\sim}{r} \in \underset{\sim}{S}$, span $R(G)$ over R. The elements $u(\underset{\sim}{r})$, $\underset{\sim}{0} \neq \underset{\sim}{r} \in \underset{\sim}{S}$, span $\Delta_R(G)$ over R.

Proof. The first assertion follows from Lemma 2.6 and the fact that every element $x \in G$ can be expressed in the canonical form (2.3). It is clear that $u(\underset{\sim}{r}) \in \Delta_R(G)$ unless $\underset{\sim}{r} = \underset{\sim}{0}$. Hence a linear combination of $u(\underset{\sim}{r})$, $\underset{\sim}{r} \in \underset{\sim}{S}$, belongs to $\Delta_R(G)$ if and only if the coefficient of $u(\underset{\sim}{0})$ (=1) is zero. This gives the second assertion of the Lemma.

The elements $u(\underset{\sim}{r})$, $\underset{\sim}{r} \in \underset{\sim}{S}$, in fact form an R-basis of $R(G)$. To prove this we first observe that if $\underset{\sim}{T} \subseteq \underset{\sim}{S}$ is a finite non-empty subset of $\underset{\sim}{S}$, then we can choose uniquely a vector $\underset{\sim}{r}^* \in \underset{\sim}{T}$ with the following property:

2.8 Given $\underset{\sim}{r}^* \neq \underset{\sim}{r} \in \underset{\sim}{T}$, let α be the first index such that $r_\alpha \neq r_\alpha^*$. Then, if $r_\alpha^* \geq 0$, we have $r_\alpha^* > r_\alpha \geq 0$ and if $r_\alpha^* < 0$, we have $r_\alpha^* < r_\alpha$.

2.9 Lemma. Let $z = \Sigma\lambda(\underset{\sim}{r})u(\underset{\sim}{r})$, $\lambda(\underset{\sim}{r}) \in R$, $\underset{\sim}{r} \in \underset{\sim}{T}$, be a finite linear combination of $u(\underset{\sim}{r})$'s. Then the coefficient of

$\Pi x_\alpha^{r_\alpha^*}$ in the expression of z as a linear combination of group elements written in the canonical form (2.3) is $\pm\lambda(\underset{\sim}{r}^*)$, where $\underset{\sim}{r}^*$ is the unique element of $\underset{\sim}{T}$ satisfying (2.8).

Proof. By definition,

$$u(\underset{\sim}{r}) = \prod_{\alpha<\lambda} v_\alpha ,$$

where

$$v_\alpha = \begin{cases} (1-x_\alpha)^{r_\alpha}, & r_\alpha \geq 0 \\ (1-x_\alpha)^M x_\alpha^{r_\alpha} & r_\alpha < 0 \end{cases}$$

Thus when we expand v_α we obtain a term $\pm x_\alpha^{r_\alpha}$ and hence $\prod_{\alpha<\lambda} x_\alpha^{r_\alpha^*}$ occurs with coefficient ± 1 in the expansion of $u(\underset{\sim}{r}^*)$. On the other hand, if $\underset{\sim}{r} \neq \underset{\sim}{r}^*$, then either $r_\alpha^* > r_\alpha \geq 0$ in which case the expansion of $(1-x_\alpha)^{r_\alpha}$ does not give a term in $x_\alpha^{r_\alpha^*}$ or $r_\alpha^* < 0$ and $r_\alpha^* < r_\alpha$ in which case neither of the possibilities for v_α gives rise to a term in $x_\alpha^{r_\alpha^*}$. This establishes Lemma 2.9.

It follows immediately from Lemma 2.9 that the elements $u(\underset{\sim}{r})$, $\underset{\sim}{r} \in \underset{\sim}{S}$, are linearly independent over R. Thus, in view of Lemma 2.7, we have

2.10 Lemma. The elements $u(\underset{\sim}{r})$, $\underset{\sim}{r} \in \underset{\sim}{S}$, are an R-basis of $R(G)$.

Let $g : N \to N$, where N is the set of natural numbers, be a map satisfying

$$g(m+n) \geq g(m) + g(n), \quad m,n \in N. \tag{2.11}$$

Suppose that the ring R is equipped with a weight function ν from $R \setminus \{0\}$ to the non-negative integers such that

$$\begin{cases} \nu(\sigma\tau) \geq \nu(\sigma) + \nu(\tau) \\ \nu(\sigma+\tau) \geq \min\{\nu(\sigma), \nu(\tau)\} . \end{cases} \tag{2.12}$$

We extend the weight function ν to the elements $u = u(\underset{\sim}{r}, \theta)$ by setting:

$$\nu(u) = \nu(\underset{\sim}{r}, \theta) = \begin{cases} \nu(\theta) + \Sigma r_\alpha g(\mu(\alpha)) & \text{if all } r_\alpha \geq 0 , \\ M & \text{otherwise.} \end{cases} \tag{2.13}$$

[see (2.2) for the definition of μ].

2.14. For $0 \leq r \leq M$, let E_r denote the R-submodule spanned by the elements $u = u(\underset{\sim}{r}, \theta)$ with $\nu(u) \geq r$. Further, let $E_r = E_M$ for all $r \geq M$.

It may be observed that the maps g = the identity map and $\nu = 0$ trivially satisfy (2.11) and (2.12). We are particularly interested in this case. In order to distinguish this case

we write $\bar{E}_r$ for the R-submodule E_r constructed using the maps g and ν given by $g(m) = m$ for all $m \in N$ and $\nu(\theta) = 0$ for $0 \neq \theta \in R$.

We note that the elements $u(\underset{\sim}{r})$ with $\nu(u(\underset{\sim}{r})) \geq r$ from an R-basis of $\bar{E}_r$ for $r = 0,1,2,\ldots,M$.

The following is the key result.

2.15 Theorem. Let $G = G_1 \supseteq G_2 \supseteq \cdots \supseteq G_c \supseteq G_{c+1} = 1$ be a finite N-series such that each of the Abelian groups G_i/G_{i+1} is a direct sum of (possibly infinitely many) cyclic groups which are either infinite or of order p^k, where p is a fixed prime and k is bounded by some integer K depending only on G. Let R be a ring with identity such that $\bigcap_i p^i R = 0$ in case G_i/G_{i+1} are not all torsion-free. Then it is possible to choose maps g and ν so that the R-submodules E_r defined as in (2.14) satisfy

$$E_r E_s \subseteq E_{r+s} \tag{2.16}$$

for all $r, s \geq 0$.

Further, if either G_i/G_{i+1} are all torsion-free or $\{G_i\}$ is a restricted N-series relative to p with $p = 0$ in R, then

$$\bar{E}_r \bar{E}_s \subseteq \bar{E}_{r+s}$$

for all $r, s \geq 0$.

Proof. Let $g : N \to N$ and $\nu : R \setminus \{0\} \to N \cup \{0\}$ be arbitrary maps satisfying (2.11) and (2.12). Let E_r be defined as in (2.14). For $\alpha < \lambda$, and $0 \leq r \leq M$, let $E_{r,\alpha}$ denote the R-submodule of $R(G)$ spanned by all elements $u = u(\underset{\sim}{r}, \theta)$ with $\nu(u) \geq r$ and $r_\beta = 0$ for $\beta \geq \alpha$. Put $E_{r,\alpha} = E_{M,\alpha}$ for $r \geq M$. The assertion (2.16) would follow if we can prove (by transfinite induction) that

$$E_{r,\alpha} E_{s,\alpha} \subseteq E_{r+s,\alpha} \tag{2.17}$$

for $\alpha < \lambda$.

First suppose $\alpha = 0$. Then $E_{r,0}$ consists of all $\theta \in R$ such that $\nu(\theta) \geq r$ together with 0. The result then follows from (2.12). If γ is a limit ordinal and x, y are elements of $E_{r,\gamma}$ and $E_{s,\gamma}$ respectively, then $x \in E_{r,\alpha}$, $y \in E_{s,\alpha}$ for a suitable $\alpha < \gamma$. Thus, if (2.17) holds for $\alpha < \gamma$, then it continues to hold with $\alpha = \gamma$.

It remains to consider the case when α has the form $\beta + 1$ for some $\beta \geq 0$.

Suppose that (2.17) holds with α replaced by any ordinal $\leq \beta$. With this inductive hypothesis, we prove the following

2.18 Proposition. Let $v = u(\underset{\sim}{r}, \theta)$ where $r_\gamma = 0$ for $\gamma > \beta$. Then

$$x_\beta^\varepsilon v \equiv v \pmod{E_{\nu(v)+\nu(\beta),\beta+1}}$$

where $\varepsilon = \pm 1$ if $x_\beta G_{\mu(\beta)+1}$ has infinite order and $\varepsilon = 1$ otherwise (here $\nu(\beta) = g(\mu(\beta))$).

The claim here is that it is possible to choose $g : N \to N$ and $\nu : R \setminus \{0\} \to N \cup \{0\}$ such that the resulting submodules $E_{r,\alpha}$ have the property of the Proposition. As we shall see, g and ν can be chosen as follows:

(a) If the factors G_i/G_{i+1} are all torsion-free, then g and ν can be any arbitrary functions satisfying (2.11) and (2.12).

(b) If at least one of the factors G_i/G_{i+1} is not torsion-free, then we can take

$$g(m) = p^{Km} \quad \text{for all} \quad m \geq 1 .$$

(c) If the series $\{G_i\}$ is a restricted N-series relative to p and $p = 0$ in R then we can take g to be any function satisfying (2.11) and, in particular, we can take $g(m) = m$ for all $m \geq 1$.

(d) If $p = 0$ in R, we can always take

$$\nu(\theta) = 0 \quad \text{for all} \quad 0 \neq \theta \in R .$$

(e) If $p \neq 0$ and at least one of G_i/G_{i+1} is not torsion-free, then we can take

$$\nu(p) = p^K g(c)$$

and

$$\nu(\theta) = n\nu(p) \quad \text{for} \quad \theta \in p^n R \setminus p^{n+1} R .$$

Proof. For $\gamma < \beta$, we have

$$\begin{aligned} x_\beta^{-1} u_\gamma x_\beta &= 1 - x_\gamma (x_\gamma, x_\beta), \text{ where } u_\gamma = 1 - x_\gamma , \\ &= z + u_\gamma (x_\gamma, x_\beta), \text{ where } z = 1 - (x_\gamma, x_\beta), \\ &= z + u_\gamma - u_\gamma z . \end{aligned}$$

Since the series $\{G_i\}$ is an N-series, $z \in \Delta_R(G_{\mu(\gamma)+\mu(\beta)})$. By Lemma 2.7, z can be expressed as a linear combination of elements $u(\underset{\sim}{s})$ with $s_\alpha = 0$ for $\alpha \geq \lambda_{\mu(\gamma)+\mu(\beta)}$. Thus, if $s_\alpha \neq 0$, then $\alpha < \beta$. Now $u(\underset{\sim}{s})$ has weight M or $\Sigma s_\alpha g(\mu(\alpha)) \geq g(\mu(\alpha))$, where α is an index chosen so that $s_\alpha > 0$. But then $\alpha < \lambda_{\mu(\gamma)+\mu(\beta)}$ and so $\mu(\alpha) \geq \mu(\gamma) + \mu(\beta)$, whence by (2.11)

$$g(\mu(\alpha)) \geq g(\mu(\gamma)) + g(\mu(\beta)) = \nu(\gamma) + \nu(\beta) .$$

Thus z is a linear combination of elements $u(\underset{\sim}{s})$ of weight $\geq \text{Min}\,\{M, \nu(\gamma)+\nu(\beta)\}$ and so $z \in E_{\nu(\gamma)+\nu(\beta),\beta}$. Now $\gamma < \beta$ and so $u_\gamma \in E_{\nu(\gamma),\beta}$. By our inductive hypothesis

$$u_\gamma z \in E_{\nu(\gamma)+\nu(\beta),\beta} .$$

Thus
$$x_\beta^{-1} u_\gamma x_\beta = u_\gamma + t \ , \ t \in E_{\nu(\gamma)+\nu(\beta),\beta}$$

and further use of our inductive hypothesis gives, for $s > 0$,

(2.19) $$x_\beta^{-1} u_\gamma^s x_\beta \equiv u_\gamma^s \pmod{E_{s\nu(\gamma)+\nu(\beta),\beta}} .$$

Also, if $\gamma < \beta$ and $s > 0$, then
$$x_\beta^{-1} u_\gamma^M x_\gamma^{-s} x_\beta = (x_\beta^{-1} u_\gamma^M x_\beta)(x_\beta^{-1} x_\gamma^{-s} x_\beta) \in E_{M,\beta}E_{0,\beta} \subseteq E_{M,\beta}$$

(by induction). Hence

(2.20) $$x_\beta^{-1} u_\gamma^M x_\gamma^{-s} x_\beta \equiv u_\gamma^M x_\gamma^{-s} \pmod{E_{M,\beta}}$$

since both sides actually belong to $E_{M,\beta}$. A similar argument shows that x_β may be replaced by x_β^{-1} in (2.19) and (2.20). Now $v = \theta v_{\alpha_1} v_{\alpha_2} \cdots v_{\alpha_k}$ where $k \geqslant 0$, $\alpha_1 < \alpha_2 < \cdots < \alpha_k \leqslant \beta$ and v_{α_i} is either $u_{\alpha_i}^{r_{\alpha_i}}$ $(r_{\alpha_j} > 0)$ or $u_{\alpha_i}^M x_{\alpha_i}^{r_{\alpha_i}}$ $(r_{\alpha_i} < 0)$. Thus, if $\alpha_k < \beta$ (we include here the case $k = 0$) , then (2.19), (2.20) and the induction hypothesis give

$$x_\beta^\varepsilon v = \theta\{\prod_{i=1}^{k} (x_\beta^\varepsilon v_{\alpha_i} x_\beta^{-\varepsilon})\} x_\beta^\varepsilon$$

(2.21) $$= \theta\{\prod_{i=1}^{k} (v_{\alpha_i} + t_i)\} x_\beta^\varepsilon \qquad (t_i \in E_{\nu(\alpha_i)+\nu(\beta),\beta})$$

(2.22) $$= (v+t) x_\beta^\varepsilon \qquad (t \in E_{\nu(v)+\nu(\beta),\beta}) .$$

By Lemma 2.6, we can write $x_\beta^\varepsilon = 1 + w$, where w is a linear combination of the elements u_β^r $(r > 0)$ and $u_\beta^M x_\beta^{-s}$ $(s > 0)$. It follows from the definition of ν and the fact that $\alpha_k < \beta$ that

$$(v+t) w \in E_{\nu(v)+\nu(\beta),\beta+1} .$$

Thus the Proposition is established in the case $\alpha_k < \beta$.

Now suppose $\alpha_k = \beta$ and $v_1 = \theta v_{\alpha_1} \cdots v_{\alpha_{k-1}}$. Then (2.22) shows that
$$x_\beta^\varepsilon v_1 = (v_1 + t_1) x_\beta^\varepsilon \qquad (t_1 \in E_{\nu(v_1)+\nu(\beta),\beta}) .$$

Hence

(2.23) $$x_\beta^\varepsilon v = (v_1+t_1)x_\beta^\varepsilon v_\beta .$$

Two cases arise depending upon whether $x_\beta G_{\mu(\beta)+1}$ has infinite or finite order.

Case I: $x_\beta G_{\mu(\beta)+1}$ has infinite order.

By Lemma 2.7,

$$R\{x_\beta\} = \sum_{r\geq 0} Ru_\beta^r + \sum_{s>0} Ru_\beta^M x_\beta^{-s} .$$

Thus $\sum_{r\geq M} Ru_\beta^r + \sum_{s>0} Ru_\beta^M x_\beta^{-s}$ is an ideal F_β, say, of $R\{x_\beta\}$ and clearly

$$E_{0,\beta}F_\beta \leq E_{M,\beta+1} .$$

Thus, if $v_\beta \in F_\beta$, then $x_\beta^\varepsilon v \in E_{M,\beta+1}$ and the Proposition is proved. Otherwise, $v_\beta = u_\beta^{r_\beta}$ $(r_\beta > 0)$ and x_β^ε is congruent mod F_β to either $1 - u_\beta$ or $1 + \sum_{\ell=1}^{M-1} u_\beta^\ell$ depending upon whether $\varepsilon = 1$ or -1 respectively. Then, mod F_β, $x_\beta^\varepsilon v_\beta$ is congruent to u^{r_β}+ a linear combination of higher powers of u_β. From (2.23) we obtain

$$x_\beta^\varepsilon v \equiv v \pmod{E_{\nu(v)+\nu(\beta),\beta+1}} .$$

This proves the Proposition in case I. It may be observed that so far we have only assumed that g and ν satisfy (2.11) and (2.12) respectively. Therefore, in case the factors G_i/G_{i+1} are all torsion-free, we can take, in particular, g = identity and $\nu(\theta) = 0$ for $0 \neq \theta \in R$.

Case II: $x_\beta G_{\mu(\beta)+1}$ has finite order p^{n_β}.

In this case $v_\beta = u_\beta^{r_\beta}$, $0 < r_\beta < p^{n_\beta}$ and $\varepsilon = 1$. Since $x_\beta = 1 - u_\beta$, we have $x_\beta v_\beta = u_\beta^{r_\beta} - u_\beta^{r_\beta+1}$ and the result follows from (2.23) unless $r_\beta = p^{n_\beta} - 1$. In this case writing $k = p^{n_\beta}$, we have

(2.24) $$u_\beta v_\beta = u_\beta^k .$$

Now
$$x_\beta^k = 1 + \sum_{i=1}^{k-1} \binom{k}{i} (-u_\beta)^i + (-u_\beta)^k .$$

Since $x_\beta^k \in G_{\mu(\beta)+1}$, it follows that

$$z = 1 - x_\beta^k \in E_{g(\mu(\beta)+1),\beta}$$

and we have

$$u_\beta^k = p \sum_{i=1}^{k-1} n_i u_\beta^i \pm z \qquad (n_i \in Z) . \tag{2.25}$$

From (2.23) and (2.24) we obtain

$$\begin{aligned} x_\beta^\varepsilon v &= (v_1+t_1)(v_\beta - u_\beta v_\beta) \\ &= v + t_1 v_\beta - (v_1+t_1)u_\beta^k . \end{aligned}$$

Now $t_1 v_\beta \in E_{\nu(v)+\nu(\beta),\beta+1}$. Using (2.25), the inductive hypothesis and the fact that $v_1 + t_1 \in E_{\nu(v_1),\beta}$, we see that $(v_1+t_1)u_\beta^k$ is a linear combination of terms which lie in $E_{\ell,\beta+1}$, where ℓ is

<u>either</u> $\qquad \nu(v_1) + g(\mu(\beta)+1)$

<u>or</u> $\qquad \nu(v_1) + \nu(p) + \nu(\beta)$.

The Proposition will, therefore, certainly hold provided that

(2.26)
$$\boxed{\nu(p) \geqslant \nu(v_\beta) = (p^{n_\beta}-1)g(\mu(\beta)) \quad \text{or} \quad p = 0 \text{ in } R}$$

and

(2.27)
$$\boxed{g(\mu(\beta)+1) \geqslant \nu(v_\beta) + \nu(\beta) = p^{n_\beta} g(\mu(\beta)) .}$$

Since $n_\beta \leqslant K$, we may put

$$g(m) = p^{Km} \quad \text{for all } m \in N .$$

Then (2.11) and (2.27) hold. If $p = 0$ in R , we put $\nu(\theta) = 0$ for $0 \neq \theta \in R$. Otherwise, let

$$\nu(p) = p^K g(c) \quad \text{(recall that } c \text{ is the length of the given N-series } \{G_i\}_{i\geq 1})$$

and if $0 \neq \theta \in R$, let

$$\nu(\theta) = i\nu(p) ,$$

where i is the largest integer such that $\theta \in p^i R$ (by hypothesis $\bigcap_i p^i R = (0)$ when the factors G_i/G_{i+1} are not all torsion-free). Then (2.12) and (2.26) hold.

Finally, <u>suppose</u> $p = 0$ in R <u>and the N-series</u> $\{G_i\}$ <u>is a restricted N-series</u>. Then each G_i/G_{i+1} is an elementary Abelian p-group. Therefore, in (2.24), $k = p$. Further, $x_\beta^k \in G_{p\mu(\beta)}$. Thus (2.25) becomes

$$u_\beta^k = z$$

and $z \in E_{pg(\mu(\beta)),\beta}$. Consequently $(v_1+t_1)u_\beta^p$ lies in $E_{\ell,\beta+1}$, where

$$\begin{aligned}\ell &= \nu(v_1) + pg(\mu(\beta)) \\ &= \nu(v_1) + \nu(v_\beta) + \nu(\beta) \\ &= \nu(v) + \nu(\beta) .\end{aligned}$$

Thus no further restriction on g is necessary. Hence we may take g to be the identity map and $\nu(\theta) = 0$ for $0 \neq \theta \in R$.

We can now complete the proof of (2.17).

By definition, $E_{r,\beta+1}$ is spanned over R by elements of the form $v = \theta v_{\alpha_1} \dots v_{\alpha_k}$ with $\alpha_1 < \dots < \alpha_k \leqslant \beta$ and $\nu(v) \geqslant r$. Suppose that $v' = \theta' v'_{\gamma_1} \dots v'_{\gamma_\ell}$ is another element with $\gamma_1 < \dots < \gamma_\ell \leqslant \beta$ and $\nu(v') \geqslant s$. If $\alpha_k < \beta$, then the inductive hypothesis gives

$$v\theta' v'_{\gamma_1} \dots v'_{\gamma_{\ell-1}} \in E_{t,\beta}$$

where $t = r + s - \nu(v'_{\gamma_\ell})$ and it is immediate that $vv' \in E_{r+s,\beta+1}$. If $\alpha_k = \beta$, we use Proposition 2.18. This gives $u_\beta v' \in E_{s+\nu(\beta),\beta+1}$. Hence we obtain, for $t > 0$, $u_\beta^t v' \in E_{s+t\nu(\beta),\beta+1}$. Also, by Proposition 2.18, $x_\beta^{-t} v' \in E_{s,\beta+1}$ for $t > 0$ whence, as above

$$u_\beta^M x_\beta^{-t} v' \in E_{M,\beta+1} .$$

It follows that

$$v_{\alpha_k} v' \in E_{s+\nu(v_{\alpha_k}),\beta+1}$$

and the result follows by the case already considered. This concludes the proof of Theorem 2.15.

Since $E_0 = R(G)$, an important consequence of the property

$$E_r E_s \subseteq E_{r+s} \quad \text{for all } r,s > 0$$

is that each E_n is an ideal of $R(G)$.

Let $\{A_n\}$ be the canonical filtration (1.5) of $R(G)$ induced by the given N-series $\{G_i\}$ (2.1). Then obviously $\Delta_R^n(G) \subseteq A_n$ for all $n \geq 1$. Suppose that $M \geq c+1$ and the R-submodules $\bar{E}_r$'s satisfy

$$\bar{E}_r\bar{E}_s \subseteq \bar{E}_{r+s} \quad \text{for all} \quad r,s \geq 0 .$$

Then each $\bar{E}_r$ is an ideal of $R(G)$. If $x \in G_i$, then $x-1$ can be written as a linear combination of $u(\underset{\sim}{r})$ with $\underset{\sim}{r} \neq \underset{\sim}{0}$, $r_\alpha = 0$ for $\alpha \geq \lambda_i$. Thus $x - 1 \in \bar{E}_i$. It follows that $A_n \subseteq \bar{E}_n$ for all $n \geq 1$. Now, by definition, $\bar{E}_n$ $(n \geq 1)$ is spanned by $u(\underset{\sim}{r})$ with $\nu(u(\underset{\sim}{r})) \geq n$. If $\nu(u(\underset{\sim}{r})) \geq n$, then

<u>either</u> all $r_{\alpha_i} \geq 0$ and $\Sigma r_{\alpha_i}\mu(\alpha_i) \geq n$

<u>or</u> some $r_{\alpha_i} < 0$.

In either case we have $u(\underset{\sim}{r}) \in A_n$ provided $n \leq M$. Hence

$$A_n = \bar{E}_n \quad \text{for} \quad n = 1,2,\ldots,M .$$

2.28 <u>Theorem</u>. <u>If</u> $\bar{E}_r\bar{E}_s \subseteq \bar{E}_{r+s}$ <u>for all</u> $r,s \geq 0$, <u>then</u>

$$A_n = \bar{E}_n \quad \underline{\text{for}} \quad n = 1,2,\ldots,M .$$

<u>Further</u>,

$$G \cap (1+\bar{E}_n) = G_n \quad \underline{\text{for all}} \quad n = 1,2,\ldots,c+1$$

<u>provided</u>

<u>either</u> R <u>is of characteristic zero</u>

<u>or</u> <u>the series</u> $\{G_i\}$ <u>is a restricted N-series relative to</u> p <u>and</u> $p = 0$ <u>in</u> R.

<u>Proof</u>. We have already seen that $A_n = \bar{E}_n$ and $G_n \subseteq 1 + \bar{E}_n$ for $n = 1,2,\ldots,c$.

Let $x \in G$ and $x - 1 \in \bar{E}_n$. Suppose $x \notin G_n$. Then there exists s such that $x \in G_s \setminus G_{s+1}$ and $1 \leq s < n$. Let $x = \prod_{\alpha<\lambda} x_\alpha^{r_\alpha}$ be the canonical expression (2.3) for x in terms of the canonical basis Φ. Then, from $x - 1 \in \bar{E}_n$, it follows that

$$\Sigma r_\alpha u_\alpha \equiv 0 \pmod{\bar{E}_{s+1}} \tag{2.29}$$

sum being taken over those α for which $\mu(\alpha) = s$. Since $\nu(u_\alpha) = s$ and $u(\underset{\sim}{r})$'s with $\nu(u(\underset{\sim}{r})) \geq s+1$ form an R-basis of $\bar{E}_{s+1}$, (2.29) implies that $r_\alpha = 0$ in R for all α with $\mu(\alpha) = s$. If R is of characteristic zero, then the integer $r_\alpha = 0$ for all α with

$\mu(\alpha) = s$. Consequently $x \in G_{s+1}$, a contradiction. If the series $\{G_i\}$ is a restricted N-series relative to a prime p and $p = 0$ in R , then each r_α is an integer satisfying $0 \leqslant r_\alpha < p$. Thus, if $r_\alpha = 0$ in R , it must be zero. Hence, once again, we get $x \in G_{s+1}$, a contradiction.

2.30 Proof of Theorem 1.6.

Let us first assume that G is finitely generated. Then G , being a finitely generated nilpotent group, satisfies the maximum condition on subgroups ([26], Lemma 1.9). Hence each H_i is finitely generated. Consequently H_i/H_{i+1} is a direct sum of cyclic groups each of which is infinite because H_i/H_{i+1} is given to be torsion-free. We can, therefore, pick a canonical basis $\Phi = (x_\alpha)_{\alpha<\lambda}$ for G . Let $\bar{E}_i$ be constructed using this basis Φ and taking $M \geqslant c + 1$. By Theorem 2.15, $\bar{E}_r\bar{E}_s \subseteq \bar{E}_{r+s}$ for all $r,s \geqslant 0$. Hence, by Theorem 2.28, $G \cap (1+A_n) = H_n$ for $n = 1,2,\ldots,c+1$. Next, let $x \in G \cap (1+A_n)$ and G be not necessarily finitely generated. Consider an expression of x as an element of $1 + A_n$, say

(2.31) $$x = 1 + a_n \ , \ a_n \in A_n \ .$$

Let G^* be the subgroup of G generated by all elements of G which appear in the equation (2.31). Then G^* is a finitely generated nilpotent group. Let

$$H_i^* = G^* \cap H_i \ .$$

It is clear that

$$G^* = H_1^* \supseteq H_2^* \supseteq \cdots \supseteq H_c^* \supseteq H_{c+1}^* = 1$$

is an N-series of G^* and that H_i^*/H_{i+1}^* is torsion-free for $i = 1,2,\ldots,c$. Thus, if A_n^* is the canonical filtration of $R(G^*)$ induced by H_i^* , then by the first part of the proof

$$H_n^* = G^* \cap (1+A_n^*) \ .$$

The equation (2.31) implies that $x \in 1 + A_n^*$. Hence $x \in H_n^* \subseteq H_n$. This completes the proof of Theorem 1.6.

2.32 Proof of Theorem 1.7.

As in the second part of the proof of Theorem 1.6, we first observe that it is enough to prove Theorem 1.7 when G is finitely generated and, therefore, a finite p-group. Let $\Phi = (x_\alpha)_{\alpha<\lambda}$ be a canonical basis for G relative to the given restricted N-series

$\{H_i\}$. Let $\bar{E}_i$ be constructed using this Φ and taking $M \geq c + 1$. By Theorem 2.15, $\bar{E}_r\bar{E}_s \subseteq \bar{E}_{r+s}$ for all $r,s \geq 0$. Hence, by Theorem 2.28,

$$G \cap (1+A_n) = H_n \quad \text{for all} \quad n = 1,2,\ldots,c+1 .$$

CHAPTER IV

DIMENSION SUBGROUPS OVER FIELDS

Let G be a group, k a field. In this Chapter we calculate the dimension subgroups $D_{n,k}(G)$ (Theorems 1.3 and 1.7) and the Lie dimension subgroups $D_{(n),k}(G)$ (Theorems 2.2 and 2.8) for all $n \geq 1$.

1. DIMENSION SUBGROUPS OVER FIELDS

Let k be a field. Theorem 2.1 of Chapter II shows that the dimension series $\{D_{n,k}(G)\}_{n\geq 1}$ of any group G depends only on the characteristic of k. More precisely, for every integer $n \geq 1$,

$$(1.1)\qquad D_{n,k}(G) = \begin{cases} D_{n,\mathbb{Q}}(G) & \text{if characteristic of } k \text{ is zero} \\ D_{n,Z/pZ}(G) & \text{if characteristic of } k \text{ is } p > 0 . \end{cases}$$

[Here $\mathbb{Q}$ denotes the field of rational numbers.]

1.2 Notation. If H is a subset of a group G, we denote by $\sqrt{H}$ the set of all elements $x \in G$ such that $x^m \in H$ for some $m > 0$.

1.3 Lemma. Let $G = \gamma_1(G) \supseteq \gamma_2(G) \supseteq \dots \supseteq \gamma_n(G) \supseteq \dots$ be the lower central series of a group G. Then $G = \sqrt{\gamma_1(G)} \supseteq \sqrt{\gamma_2(G)} \supseteq \dots \sqrt{\gamma_n(G)} \supseteq \dots$ is an N-series in G.

Proof. Cleary $\gamma_n(G) \subseteq \sqrt{\gamma_n(G)}$. Since the periodic elements of a nilpotent group form a subgroup, $\sqrt{\gamma_n(G)}$ is a subgroup of G and is trivially normal. Let $x,y \in G$ and $r,s > 0$ be integers such that $x^r \in \gamma_m(G)$ and $y^s \in \gamma_n(G)$. To show that the commutator $(x,y) \in \sqrt{\gamma_{n+m}(G)}$, we can assume that $\sqrt{\gamma_{n+m}(G)} = 1$. Suppose $(x,y) \neq 1$. Then there exists an integer $i > 1$ such that $(x,y) \in Z_i \setminus Z_{i-1}$, where $\{Z_i\}_{i\geq 0}$ is the upper central series of G. For, the assumption $\sqrt{\gamma_{n+m}(G)} = 1$ makes G a (torsion-free) nilpotent group. Now

$$1 = (x^r, y^s) \equiv (x,y)^{rs} \pmod{Z_{i-1}}$$

which implies that $(x,y)Z_{i-1}$ is a non-identity torsion element of G/Z_{i-1}. This, however, is not possible, since G is torsion-free nilpotent. Hence $(x,y) = 1$ and $\{\sqrt{\gamma_n(G)}\}_{n\geq 1}$ is an N-series.

1.4 Lemma. The canonical filtration $\{A_n\}_{n\geq 1}$ of $\Delta_{\mathbb{Q}}(G)$ induced by the N-series $\{\sqrt{\gamma_n(G)}\}_{n\geq 1}$ is the $\Delta_{\mathbb{Q}}(G)$-adic filtration, i.e.

$$A_n = \Delta_{\mathbb{Q}}^n(G)$$

for all $n \geq 1$.

Proof. By definition, A_n is the $\mathbb{Q}$-subspace of $\mathbb{Q}(G)$ spanned by the products $(g_1-1)(g_2-1)\ldots(g_s-1)$, $(s\geq 1)$, with $\sum_j w(g_j) \geq n$, where w is the weight function induced on G by the N-series $\{\sqrt{\gamma_n(G)}\}_{n\geq 1}$. Let $g \in \sqrt{\gamma_n(G)}$. If $g^m \in \gamma_n(G)$, $m > 0$, then $g^m - 1 \in \Delta_{\mathbb{Q}}^n(G)$ and the equation $g^m - 1 = m(g-1) + \binom{m}{2}(g-1)^2 + \ldots + (g-1)^m$ shows that $g - 1 \in \Delta_{\mathbb{Q}}^n(G)$. It follows that $\sqrt{\gamma_n(G)} \subseteq D_{n,\mathbb{Q}}(G)$ and $A_n \subseteq \Delta_{\mathbb{Q}}^n(G)$ for all $n \geq 1$. As $\Delta_{\mathbb{Q}}^n(G) \subseteq A_n$ for all $n \geq 1$, we have $A_n = \Delta_{\mathbb{Q}}^n(G)$ for all $n \geq 1$.

1.5 Theorem ([26], [36]). For all $n \geq 1$,

$$D_{n,\mathbb{Q}}(G) = \sqrt{\gamma_n(G)} .$$

Proof. Let $n \geq 1$ be given. To show that $\sqrt{\gamma_n(G)} = D_{n,\mathbb{Q}}(G)$ we can assume, by going mod $\sqrt{\gamma_n(G)}$ if necessary, that $\sqrt{\gamma_n(G)} = 1$. Then $\sqrt{\gamma_i(G)}/\sqrt{\gamma_{i+1}(G)}$ being torsion-free for all $i \geq 1$, Theorem 1.6 of Chapter III implies that $G \cap (1+\Delta_{\mathbb{Q}}^n(G)) = G \cap (1+A_n) = \sqrt{\gamma_n(G)}$, where $\{A_n\}_{n\geq 1}$ is the canonical filtration of $\Delta_{\mathbb{Q}}(G)$ induced by the N-series $\{\sqrt{\gamma_i(G)}\}_{i\geq 1}$ (Lemmas 1.3 and 1.4).

We next proceed to calculate the dimension series $\{D_{n,Z/pZ}(G)\}_{n\geq 1}$ where G is an arbitrary group and p is a prime. By (Chapter III, 1.3) this series is a restricted N-series relative to p. Thus, in particular, it is a central series and has the property:

(1.6) $\quad x \in D_{i,Z/pZ}(G) \Rightarrow x^p \in D_{ip,Z/pZ}(G)$ for all $i \geq 1$.

Therefore, the series $\{D_{n,Z/pZ}(G)\}_{n\geq 1}$ must contain the Brauer-Jennings-Zassenhaus M-series $\{M_{n,p}(G)\}_{n\geq 1}$ which is the minimal central series with the property:

(1.7) $\quad x \in M_{n,p}(G) \Rightarrow x^p \in M_{np,p}(G)$ for all $n \geq 1$.

The series $\{M_{n,p}(G)\}_{n\geq 1}$ is defined inductively by

$$M_{1,p}(G) = G , \; M_{n,p}(G) = (G, M_{n-1,p}(G))M^p_{(\frac{n}{p})}(G)$$

for $n \geq 2$, where $(\frac{n}{p})$ is the least integer $\geq \frac{n}{p}$. As $\{M_{n,p}(G)\}_{n\geq 1}$ is a central series, it contains the lower central series $\{\gamma_n(G)\}_{n\geq 1}$ of G . In view of (1.7) it follows that

(1.8) $$x \in \gamma_i(G) \;,\; ip^j \geq n \Rightarrow x^{p^j} \in M_{n,p}(G) \;.$$

Hence the M-series contains the series $\{G_{n,p}\}_{n\geq 1}$ of normal subgroups defined by setting

$$G_{n,p} = \prod_{ip^j\geq n} \gamma_i(G)^{p^j} \;,\; n \geq 1$$

1.9 <u>Theorem</u> ([7], [35], [38], [80], [99]). <u>For every group</u> G , <u>prime</u> p <u>and integer</u> $n \geq 1$,

$$G_{n,p} = M_{n,p}(G) = D_{n,Z/pZ}(G) \;.$$

We plan to apply Theorem 1.7 of Chapter III to prove the above Theorem. For this purpose we need to check that $\{G_{n,p}\}_{n\geq 1}$ is a restricted N-series of G . The following theorem of Dark [15] is very useful in this context. [Alternatively, one can follow Lazard [38] and first prove Theorem 1.9 for free groups from which it follows immediately that, for any group G , $\{G_{n,p}\}_{n\geq 1}$ is restricted N-series.] We are thankful to Dr. R. Dark for allowing his theorem and its proof to be included here.

1.10 <u>Definitions</u>. The <u>components</u> of a commutator are the group elements which appear in it. An <u>X-commutator</u> is a commutator with all its components in X . An <u>X-product</u> is a product with all its factors in X . $X^{-1} = \{x^{-1} : x \in X\}$, and $X^{\pm 1} = X \cup X^{-1}$. A binomial coefficient $\binom{u}{v}$ is defined to be 0 unless $0 \leq v \leq u$. An <u>m-tuple</u>

$(r) = (r_1,\ldots,r_m)$ is an m-tuple of non-negative integers;
$(0) = (0,\ldots,0)$. If (r) and (s) are m-tuples, then

$$\binom{r}{s} = \binom{r_1}{s_1}\binom{r_2}{s_2}\cdots\binom{r_m}{s_m} \;.$$

We order the m-tuples (r) using dictionary order for the m-tuples $(r_1+\ldots+r_m \;,\; r_1,r_2 \;,\ldots,\; r_{m-1})$; in any product indexed by m-tuples, the factors are taken in this order. If x is a group element, then $x^0 = 1$.

1.11 <u>Theorem</u> [15]. <u>Let</u> m <u>be a positive integer, and suppose that</u> $Y_1,\ldots,Y_m$ <u>are disjoint subsets of some group. Take</u> $Y = Y_1 \cup\ldots\cup Y_m$

and let π be any Y-product. If (r) is an m-tuple, we put $\Lambda(r)$ for the set of $Y^{\pm 1}$-commutators χ with the property that, for each $\alpha \leq m$, χ has at least r_α components in $Y_\alpha^{\pm 1}$, and we write $\pi(r)$ for the expression obtained from π by replacing each factor y in $Y_\alpha^{\pm 1}$ by y^{r_α} $(1 \leq \alpha \leq m)$. If $\theta(r)$ is defined inductively by the equations $\theta(0) = 1$,

$$\theta(r) = \left[\prod_{(s)<(r)} \theta(s)^{\binom{r}{s}} \right]^{-1} \pi(r) \text{ when } (r) > (0)$$

then $\theta(r)$ is equal to a $\Lambda(r)$-product.

Proof. We remark first that if the result holds in the group D freely generated by Y, then it is true a fortiori in any group of which Y is a subset; we shall therefore work in D. Since π, regarded as a Y-product, has only a finite number of factors, we may assume that Y is finite. Suppose that $Y = \{y_1,\ldots,y_u\}$, and put $U = \{1,\ldots,u\}$, $U_\alpha = \{i \leq u: y_i \in Y_\alpha\}$ $(1 \leq \alpha \leq m)$. We write $v = r_1 + \ldots + r_m$ and $V = \{1,\ldots,v\}$, and we choose disjoint subsets $V_1,\ldots,V_m$ of V with $|V_\alpha| = r_\alpha$ $(1 \leq \alpha \leq m)$. Then $U = U_1 \cup \ldots \cup U_m$ and $V = V_1 \cup \ldots \cup V_m$. Take distinct symbols z_{ij}, and let E be the group freely generated by the set $Z = \{z_{ij}: i \in U_\alpha, j \in V_\alpha, 1 \leq \alpha \leq m\}$. We define a homomorphism $f: D \to E$ by taking

$$f(y_i) = \prod_{j \in V_\alpha} z_{ij} \quad \text{when } i \in U_\alpha \quad (1 \leq \alpha \leq m),$$

where the order of the factors on the right hand side is given by the natural ordering of V_α (which is a set of integers).

We define the support of z_{ij} to be $\{j\}$, and the support of any $Z^{\pm 1}$-commutator χ to be the union of the supports of its components. Thus the support of χ is a subset of V; if S is any subset of V, we write $\Lambda(S)$ for the set of $Z^{\pm 1}$-commutators whose support is S. We denote the m-tuple $(|S \cap V_1|,\ldots,|S \cap V_m|)$ by $(|S \cap V_\alpha|)$, and we choose an ordering for the subsets of V in such a way that S precedes T whenever $(|S \cap V_\alpha|) < (|T \cap V_\alpha|)$ in our ordering of m-tuples; any product indexed by subsets of V will be taken in this order. We note that if $S \subset T \subseteq V$, then S precedes T.

The first step in the proof of the Theorem is to find an expression for the Z-product $f(\pi)$ of the form

$$(1.12) \qquad f(\pi) = \prod_{S \subseteq V} \theta(S)$$

where each factor $\theta(S)$ is a $\Lambda(S)$-product (see [32] Chapter III, 9.3). This is done by applying to $f(\pi)$ a 'collecting process'. If S is the first subset of V in our ordering, we begin by moving all the factors in $\Lambda(S)$ to the left to form $\theta(S)$, introducing commutators as we do so; we then do the same for the second subset of V, and so on. If S and T are subsets of V, and S precedes T, and if $\lambda \in \Lambda(S)$ and $\mu \in \Lambda(T)$, then a typical step consists of replacing a segment $\mu\lambda$ by $\lambda\mu(\mu,\lambda)$. It is not hard to see that $S \subset S \cup T$, and so S precedes $S \cup T$; since $(\mu,\lambda) \in \Lambda(S \cup T)$, this means that we do not introduce commutators which should already have been collected. It follows that the process ends after a finite number of steps and we obtain the expression (1.12).

We define a homomorphism $g : E \to D$ by taking

$$g(z_{ij}) = y_i \quad (z_{ij} \in Z) .$$

Let S be a subset of V, and put $(s) = (|S \cap V_\alpha|)$.
We shall prove that

$$g(\theta(S)) = \theta(s) , \tag{1.13}$$

where $\theta(s)$ is as defined in the statement of the Theorem. If S is the empty set, then $\theta(S) = 1$ and $\theta(s) = \theta(0) = 1$. We may therefore assume that $|S| > 0$, and use induction on $|S|$.

We define an endomorphism e of E by taking

$$e(z_{ij}) = \begin{cases} z_{ij} & \text{when } j \in S , \\ 1 & \text{when } j \notin S . \end{cases}$$

If $T \subseteq S$, then the components of any commutator in $\Lambda(T)$ all lie in the set $\{z_{ij}^{\pm 1} : z_{ij} \in Z , j \in S\}$, and so are fixed by e; since $\theta(T)$ is a $\Lambda(T)$-product, it follows that $e(\theta(T)) = \theta(T)$. On the other hand if $T \subseteq V$ but $T \not\subseteq S$, and if $\chi \in \Lambda(T)$, then at least one component of χ is sent to 1 by e, and so $e(\chi) = 1$; hence $e(\theta(T)) = 1$ in this case. Applying e to (1.12), we therefore get

$$e(f(\pi)) = \prod_{T \subseteq S} \theta(T) .$$

It follows that if we define

$$\varphi = \prod_{T \subset S} g(\theta(T)) ,$$

then

$$g(e(f(\pi))) = \varphi\, g(\theta(S) . \tag{1.14}$$

We now examine the product φ. If $T \subset S$ and $(|T \cap V_\alpha|) = (t)$ then, since we are proving (1.13) by induction we may assume that $g(\theta(T)) = \theta(t)$. Moreover, for each $(t) < (s)$, the number of subsets T of S such that $(|T \cap V_\alpha|) = (t)$ is

$$\binom{s_1}{t_1}\binom{s_2}{t_2}\cdots\binom{s_m}{t_m} = \binom{s}{t} ;$$

this is true even when $t_\alpha > s_\alpha$ for some α, because of our convention for binomial coefficients. Finally, we remark that our choice of ordering for the subsets of V ensures that all the factors of φ which are equal to $\theta(t)$ occur together. Hence the definition of φ can be rewritten as

$$(1.15) \qquad \varphi = \prod_{(t)<(s)} \theta(t)^{\binom{s}{t}} .$$

We next consider $g(e(f(\pi)))$. If $1 \leqslant \alpha \leqslant m$ and $i \in U_\alpha$, then

$$g(e(f(y_i))) = \prod_{j\in V_\alpha} g(e(z_{ij})) = \prod_{j\in S\cap V_\alpha} g(z_{ij}) = y_i^{|S\cap V_\alpha|}$$
$$= y_i^{s_\alpha} .$$

It follows that $g(e(f(\pi))) = \pi(s)$. Combining this with (1.14) and (1.15), we get

$$g(\theta(S)) = \varphi^{-1} g(e(f(\pi))) = \left[\prod_{(t)<(s)} \theta(t)^{\binom{s}{t}}\right]^{-1} \pi(s) .$$

This is the same as the definition of $\theta(s)$, and we have therefore proved (1.13). In particular $\theta(r) = g(\theta(V))$, which is a $g(\Lambda(V))$-product. Since $g(\Lambda(V)) = \Lambda(r)$, this proves the Theorem.

An application of Dark's Theorem to the word $\pi = (x,y)$ gives

1.16 Corollary. If g_1, g_2 are elements of a group G and α_1, α_2 are integers $\geqslant 1$, then

$$(g_1^{\alpha_1}, g_2^{\alpha_2}) = \Pi\theta(r_1,r_2)^{\binom{\alpha_1}{r_1}\binom{\alpha_2}{r_2}}$$

where the product is taken over all pairs (r_1,r_2) with $1 \leqslant r_i \leqslant \alpha_i$ and where $\theta(r_1,r_2)$ is a product of $\{g_1,g_2\}$-commutators each containing at least r_i components equal to g_i or g_i^{-1}, $i = 1,2$.

It is the Corollary 1.16 that we need to verify that certain series are N-series.

Let N denote the set of natural numbers, $N_o = N \cup \{0\}$, $N_\infty = N \cup \{\infty\}$. If $F(i,\alpha) : N \times N_o \to N_\infty$ is a map satisfying $F(i,0) = \infty$ for all i and G is a group, then we can define a decreasing series $\{G^F(r)\}_{r\geq 1}$ of normal subgroups of G by setting

$$(1.17) \qquad G^F(r) = \prod_{F(i,\alpha)\geq r} \gamma_i(G)^\alpha$$

where $\{\gamma_i(G)\}_{i\geq 1}$ is the lower central series of G.

1.18 Corollary. (i) Let $F : N \times N_o \to N_\infty$ be a map satisfying

$$(1.19) \qquad F(ri+sj, \tbinom{\alpha}{r}\tbinom{\beta}{s}) \geq F(i,\alpha) + F(j,\beta)$$

for all r, s, $1 \leq r \leq \alpha$, $1 \leq s \leq \beta$. Then $\{G^F(r)\}_{r\geq 1}$ is an N-series of $G^F(1)$.

(ii) Let p be a prime and suppose that in addition to (1.19) F also satisfies the following:

$$(1.20) \qquad F(i,p\alpha) \geq pF(i,\alpha) \quad \text{for all} \quad i,\alpha \in N.$$

Then $\{G^F(r)\}_{r\geq 1}$ is a restricted N-series of $G^F(1)$ relative to the prime p.

Proof. (i) To show that $\{G^F(r)\}_{r\geq 1}$ is an N-series, it is clearly enough to prove that if $g_1 \in \gamma_i(G)$, $g_2 \in \gamma_j(G)$, and α,β are integers ≥ 1 satisfying

$$F(i,\alpha) \geq m \quad \text{and} \quad F(j,\beta) \geq n,$$

then

$$(g_1^\alpha, g_2^\beta) \in G^F(m+n).$$

By Corollary 1.16, we have

$$(g_1^\alpha, g_2^\beta) = \Pi\theta(r,s)^{\binom{\alpha}{r}\binom{\beta}{s}}$$

where the product is taken over all pairs (r,s) with $1 \leq r \leq \alpha$, $1 \leq s \leq \beta$ and $\theta(r,s) \in \gamma_{ri+sj}(G)$. Since F satisfies (1.19),

$\theta(r,s)^{\binom{\alpha}{r}\binom{\beta}{s}} \in G^F(m+n)$. Hence $(g_1^\alpha, g_2^\beta) \in G^F(m+n)$.

(ii) Now suppose F satisfies the hypothesis of (ii). Then by (i) $\gamma_p(G^F(r)) \subseteq G^F(pr)$ for all $r \geq 1$. Thus each of the factors $G^F(r)/G^F(pr)$ is a nilpotent group of class less than p. Let $g \in \gamma_i(G)$ and $\alpha \in N$ be such that $F(i,\alpha) \geq r$. Then, by hypothesis, $F(i,p\alpha) \geq pr$. Thus $g^{p\alpha} \in G^F(pr)$. Hence $G^F(r)/G^F(pr)$ is generated by elements of order p. Consequently $G^F(r)/G^F(pr)$ has period p, i.e. $G^F(r)^p \subseteq G^F(pr)$ for all $r \geq 1$ (see [75], Chapter XI, Lemma 1.15).

1.21 <u>Notation</u>. If p is a prime and n is a natural number, then we write

$$p^j \| n$$

to mean that p^j <u>divides</u> n and p^{j+1} does <u>not</u> divide n.

1.22 <u>Lemma</u> [80]. <u>The series</u> $\{G_{n,p}\}_{n\geq 1}$ <u>is a restricted N-series of</u> G <u>relative to the prime</u> p.

<u>Proof</u>. Let $F : N \times N_0 \to N_\infty$ be defined by

$$F(i,\alpha) = ip^j$$

where $p^j \| \alpha, \alpha \neq 0$, and

$$F(i,0) = \infty .$$

We assert that F satisfies the conditions (1.19) and (1.20) of Corollary 1.12. Let $\alpha = p^k\alpha'$, $\beta = p^\ell\beta'$, $\alpha,\beta,\alpha',\beta' \in N$, $p \nmid \alpha'\beta'$. Let $r,s \in N$, $1 \leq r \leq \alpha$, $1 \leq s \leq \beta$, $r = p^a b$, $s = p^c d$, $a,b,c,d \in N$, $p \nmid bd$. If $i,j \in N$, then $F(ri+sj, \binom{\alpha}{r}\binom{\beta}{s}) = (ri+sj)p^{k-a+\ell-c}$, because $p^{k-a} \| \binom{\alpha}{r}$ and $p^{\ell-c} \| \binom{\beta}{s}$. Now

$$\begin{aligned}(ri+sj)p^{k+\ell-a-c} &= (ip^a b+jp^c d)p^{k+\ell-a-c} \\ &= ip^k bp^{\ell-c} + jp^\ell dp^{k-a} \\ &\geq ip^k + jp^\ell \qquad (\text{since } \ell \geq c,\ k \geq a) \\ &= F(i,\alpha) + F(j,\beta) .\end{aligned}$$

Hence F satisfies (1.19). That F satisfies (1.20) is trivial. Hence, by Corollary 1.18, $\{G^F(r)\}_{r\geq 1}$ is a restricted N-series of $G^F(1) = G$. As $G^F(r) = G_{r,p}$, the Lemma is proved.

1.23 <u>Lemma</u>. <u>The canonical filtration</u> $\{A_n\}_{n\geq 1}$ <u>of</u> $\Delta_{Z/pZ}(G)$ <u>induced by the N-series</u> $\{G_{n,p}\}_{n\geq 1}$ <u>is the</u> $\Delta_{Z/pZ}(G)$<u>-adic filtration, i.e.</u>

$$A_n = \Delta^n_{Z/pZ}(G)$$

<u>for all</u> $n \geq 1$.

<u>Proof</u>. Since $\Delta^n_{Z/pZ}(G) \subseteq A_n$ for all $n \geq 1$, it suffices to prove that $G_{n,p} \subseteq 1 + \Delta^n_{Z/pZ}(G)$ for all $n \geq 1$. However, we have already observed that

$$G_{n,p} \subseteq D_{n,p}(G) \quad \text{for all} \quad n \geq 1$$

[See the remarks preceding the statement of Theorem 1.9]. Hence the canonical filtration of $\Delta_{Z/pZ}(G)$ induced by the N-series $\{G_{n,p}\}_{n\geq 1}$

is the $\Delta_{Z/pZ}(G)$-adic filtration.

1.24 Proof of Theorem 1.9.

Let n be an integer ≥ 1 . Then $G_{n,p} \subseteq D_{n,Z/pZ}(G)$. Thus, in order to show that $G_{n,p} = D_{n,Z/pZ}(G)$, we may assume that $G_{n,p} = 1$ and then show that $D_{n,Z/pZ}(G)$ must also be 1 . Let $x \in 1 + \Delta^n_{Z/pZ}(G)$. By considering an expression for x as an element of $1 + \Delta^n_{Z/pZ}(G)$, it follows that we may assume G to be finitely generated. But then, in view of the assumption $G_{n,p} = 1$, G is a finite p-group. Thus it is enough to prove that for a finite p-group G with $G_{n,p} = 1$, $D_{n,Z/pZ}(G) = 1$. By Lemma 1.23, the canonical filtration of $\Delta_{Z/pZ}(G)$ induced by $\{G_{i,p}\}_{i\geq 1}$ is the $\Delta_{Z/pZ}(G)$-adic filtration of $\Delta_{Z/pZ}(G)$. Hence, by Theorem 1.7 of Chapter III,

$$D_{n,Z/pZ}(G) = G \cap (1+\Delta^n_{Z/pZ}(G)) = G_{n,p} = 1 .$$

2. LIE DIMENSION SUBGROUPS OVER FIELDS

Let k be a field, G a group. Then, by Theorem 3.1 of Chapter II, for all $n \geq 1$

$$D_{(n),k}(G) = \begin{cases} D_{(n)\mathbb{Q}}(G) & \text{if characteristic of } k \text{ is zero,} \\ D_{(n),Z/pZ}(G) & \text{if characteristic of } k \text{ is } p > 0 . \end{cases}$$

2.1 Proposition. *If G is a group and R a commutative ring with identity, then the Lie dimension subgroups* $D_{(n),R}(G) = G \cap (1+\Delta_R^{(n)}(G))$, $n \geq 1$, *have the following properties:*

(i) $$D_{(2),R}(G) = \gamma_2(G)$$

where $\gamma_2(G)$ *is the derived group of* G .

(ii) $$(D_{(m),R}(G), D_{(n),R}(G)) \subseteq D_{(m+n),R}(G) \quad \text{for all } m,n \geq 1 ,$$

i.e. $\{D_{(n),R}(G)\}_{n\geq 1}$ *is an N-series*.

Proof (i) Let $x,y \in G$. Then

$$(x,y) - 1 = x^{-1}y^{-1}\{(x-1)(y-1) - (y-1)(x-1)\}$$
$$\in \Delta_R^{(2)}(G) .$$

Hence $\gamma_2(G) \subseteq D_{(2),R}(G)$. For the reverse inclusion we may assume that G is Abelian. But then $\Delta_R^{(2)}(G) = 0$ and so $D_{(2),R}(G) = 1$.

(ii) If $x \in D_{(m),R}(G)$ and $y \in D_{(n),R}(G)$, then $x - 1 \in \Delta_R^{(m)}(G)$ and $y - 1 \in \Delta_R^{(n)}(G)$. Therefore, $(x,y) - 1 = x^{-1}y^{-1}[x-1,y-1] \in \Delta_R^{(m+n)}(G)$ (Chapter I, Proposition 1.7). Hence $(x,y) \in D_{(m+n),R}(G)$.

The Lie dimension subgroups over $\mathbb{Q}$ are easy to calculate and are given by the following

2.2 Theorem [71]. For all $n \geq 2$

$$D_{(n),\mathbb{Q}}(G) = \sqrt{\gamma_n(G)} \cap \gamma_2(G)$$

where $\{\gamma_n(G)\}_{n\geq 1}$ is the lower central series of G.

Proof. Since $D_{(n),\mathbb{Q}}(G) \subseteq D_{n,\mathbb{Q}}(G)$ and $D_{(2),\mathbb{Q}}(G) = \gamma_2(G)$ (Proposition 2.1), it follows from Theorem 1.5 that $D_{(n),\mathbb{Q}}(G) \subseteq \sqrt{\gamma_n(G)} \cap \gamma_2(G)$ for all $n \geq 2$.

Conversely, let $x \in \sqrt{\gamma_n(G)} \cap \gamma_2(G)$ for some $n \geq 2$. Then there exists $m > 0$ such that $x^m \in \gamma_n(G)$. Consider the equation

$$(*) \qquad x^m - 1 = m(x-1) + \binom{m}{2}(x-1)^2 + \ldots + (x-1)^m.$$

Suppose $x - 1 \in \Delta_{\mathbb{Q}}^{(j)}(G)$, $x - 1 \notin \Delta_{\mathbb{Q}}^{(j+1)}(G)$ and $2 \leq j < n$. Then (*) shows that

$$x - 1 \in \Delta_{\mathbb{Q}}^{(s)}(G), \quad s = \min(n, 2j-1)$$

by virtue of (Chapter I, 1.7 (iii)). Since $s \geq j + 1$, we have a contradiction. Hence $x - 1 \in \Delta_{\mathbb{Q}}^{(n)}(G)$ and so $x \in D_{(n),\mathbb{Q}}(G)$.

2.3 Definitions (i). Let G be a group, p a prime. We define a series $\{M_{(n),p}(G)\}_{n\geq 1}$ by setting:

$$M_{(1),p}(G) = G, \quad M_{(2),p}(G) = \gamma_2(G), \quad M_{(n+1),p}(G) = (G, M_{(n),p}(G))\, M^p_{(\frac{n+p}{p})}(G)$$

for $n \geq 2$, where, for a rational number $\frac{r}{s}$, $(\frac{r}{s})$ denotes the least integer $\geq \frac{r}{s}$.

The series $\{M_{(n),p}(G)\}_{n\geq 1}$ is the smallest descending central series with the property

$$(2.4) \qquad M^p_{(i),p}(G) \subseteq M_{(ip-p+1),p}(G) \quad \text{for all } i \geq 1.$$

(ii) We define a series $\{G_{(n),p}\}_{n\geq 1}$ by setting

$$G_{(n),p} = \prod_{(i-1)p^j \geq n} \gamma_i(G)^{p^j}.$$

Note that $G_{(1),p} = \gamma_2(G)$.

2.5 Lemma [71]. *Let* G *be a group,* p *a prime. Then the series* $\{G_{(n),p}\}_{n\geq 1}$ *is a restricted N-series of* $\gamma_2(G)$.

Proof. Let $F : N \times N_o \to N_\infty$ be defined by

$$F(i,\alpha) = (i-1)p^j$$

where $p^j \| \alpha$, $\alpha \neq 0$ and

$$F(i,0) = \infty.$$

Then a verification similar to the one carried out in the proof of Lemma 1.22 shows that F satisfies (1.19) and (1.20). Hence, by Corollary 1.18, $\{G^F(r)\}_{r\geq 1}$ is a restricted N-series of $G^F(1) = \gamma_2(G)$. As $G^F(r) = G_{(r),p}$, the Lemma is proved.

2.6 Lemma *For every group* G *and prime* p

$$G_{(n),p} \subseteq M_{(n+1),p}(G) \subseteq D_{(n+1),Z/pZ}(G)$$

for all $n \geq 1$.

Proof. Since $\{M_{(n),p}(G)\}_{n\geq 1}$ is a central series in G, $\gamma_i(G) \subseteq M_{(i),p}(G)$ for all $i \geq 1$. In view of the property (2.4) of the series $\{M_{(n),p}(G)\}_{n\geq 1}$, we have

$$\gamma_i(G)^{p^j} \subseteq M_{((i-1)p^j+1),p}(G) .$$

Thus, if $ip^j \geq n + p^j$, then $\gamma_i(G)^{p^j} \subseteq M_{(n+1),p}(G)$. Hence $G_{(n),p} \subseteq M_{(n+1),p}(G)$ for all $n \geq 1$.

We prove the second inclusion by induction on $n \geq 1$. Since $M_{(2),p}(G) = \gamma_2(G) = D_{(2),Z/pZ}(G)$, the inclusion holds for $n = 1$. Suppose $m \geq 2$ and the inclusion holds for all $n < m$. Then

$$M_{(m+1),p}(G) = (G, M_{(m),p}(G))M^p_{(\frac{m+p}{p}),p}(G).$$

By induction,

$$M_{(m),p}(G) \subseteq D_{(m),Z/pZ}(G) .$$

Hence

$(G,M_{(m),p}(G)) \subseteq (G,D_{(m),Z/pZ}(G)) \subseteq D_{(m+1),Z/pZ}(G)$. (Proposition 2.1 (ii)).

Let $s = (\frac{m+p}{p})$ and $x \in M_{(s),p}(G)$. Then, by induction, $x \in D_{(s),Z/pZ}(G)$, i.e. $x - 1 \in \Delta^{(s)}_{Z/pZ}(G)$. Hence

$$x^p - 1 = (x-1)^p \in (\Delta^{(s)}_{Z/pZ}(G))^p \subseteq \Delta^{(ps-p+1)}_{Z/pZ}(G)$$

(Chapter I, 1.7 (iii)). Since $s \geq \frac{m+p}{p}$, $ps - p + 1 \geq m + 1$. Consequently, $x^p - 1 \in \Delta^{(m+1)}_{Z/pZ}(G)$ and so $x^p \in D_{(m+1),Z/pZ}(G)$. Thus we have shown that $M_{(m+1),p}(G) \subseteq D_{(m+1),Z/pZ}(G)$. This completes the induction and so establishes the second inclusion of the Lemma for all $n \geq 1$.

2.7 Lemma. *Let R be a commutative ring with identity, G a group. If the characteristic of R is p, a prime, then the canonical filtration $\{A_n\}_{n\geq 1}$ of $R(\gamma_2(G))$ defined by the restricted N-series $\{G_{(n),p}\}_{n\geq 1}$ is given by*

$$A_n = R(\gamma_2(G)) \cap \Delta_R^{(n+1)}(G) \quad \text{for all} \quad n \geq 1.$$

Further, $A_n \cdot R(G) = \Delta_R^{(n+1)}(G)$ *for all* $n \geq 1$.

Proof. By definition, $\gamma_{i+1}(G) \subseteq G_{(i),p}$ for all $i \geq 1$. Let $g_i \in G_{(n_i),p}$ for $i = 1,2,\ldots,r$. Then, $g_i \in \gamma_2(G)$ and, by Lemma 2.6, $g_i - 1 \in \Delta^{(n_i+1)}_{Z/pZ}(G) \subseteq \Delta^{(n_i+1)}_R(G)$. Therefore,

$$(g_1-1)(g_2-1)\ldots(g_r-1) \in \Delta_R^{(n_1+1)}(G)\ldots\Delta_R^{(n_r+1)}(G)$$
$$\subseteq \Delta_R^{(n_1+n_2+\ldots+n_r+1)}(G) \quad \text{(Chapter I, 1.7 (iii)).}$$

Thus, if $\sum_{i=1}^{r} n_i \geq n$, then $(g_1-1)(g_2-1)\ldots(g_r-1) \in \Delta_R^{(n+1)}(G)$. Hence $A_n \subseteq R(\gamma_2(G)) \cap \Delta_R^{(n+1)}(G)$.

Conversely, let $x \in R(\gamma_2(G)) \cap \Delta_R^{(n+1)}(G)$. Then, by (Chapter I, 1.8)

(*) $$x = x_{n+1} + \Sigma \prod_j x_{n_j} y \;,\; y \in G \;,\; x_{n+1} \in \Delta_R(G,\gamma_{n+1}(G)) \;,$$

$x_{n_j} \in \Delta_R(\gamma_{n_j}(G))$, $\Sigma(n_j-1) = n$. Choose a transversal T of $\gamma_2(G)$ in G and represent every element $g \in G$ as $g = ht$, $h \in \gamma_2(G)$, $t \in T$. Then we have an R-homomorphism $\alpha : R(G) \to R(\gamma_2(G))$ induced by $g \to h$. An application of α to the equation (*) shows that $x \in A_n$.

The last assertion is clear from Theorem 1.8 of Chapter I.

We are now ready to give the structure of the Lie dimension subgroups $D_{(n),Z/pZ}(G)$, $n \geq 1$.

2.8 Theorem [71]. *For every group* G *and prime* p ,

$$G_{(n),p} = M_{(n+1),p}(G) = D_{(n+1),Z/pZ}(G)$$

for all $n \geq 1$.

Proof. Let $n \geq 1$. By Lemma 2.6 $G_{(n),p} \subseteq M_{(n+1),p}(G) \subseteq D_{(n+1),Z/pZ}(G)$. Thus to establish the reverse inclusions, we can assume, by going mod $G_{(n),p}$ if necessary, that $G_{(n),p} = 1$. Then G is nilpotent and $\gamma_2(G)$ is of exponent a power of p . It is clear that to prove $D_{(n+1),Z/pZ}(G) = 1$ we may assume further that G is finitely generated. Then $\gamma_2(G)$ is a finitely generated nilpotent group of p power exponent and, therefore, has order a power of p . Now $\{G_{(i),p}\}_{i\geq 1}$ is a restricted N-series of $\gamma_2(G)$ (Lemma 2.5) and, by Lemma 2.7, the canonical filtration $\{A_i\}_{i\geq 1}$ of $(Z/pZ)(\gamma_2(G))$ defined by this series is given by

$$A_i = (Z/pZ)(\gamma_2(G)) \cap \Delta^{(i+1)}_{Z/pZ}(G) \quad \text{for all } i \geq 1 .$$

Therefore, by (Chapter I, Theorem 1.7),

$$\begin{aligned} 1 = G_{(n),p} &= \gamma_2(G) \cap (1+A_n) \\ &= \gamma_2(G) \cap (1+\Delta^{(n+1)}_{Z/pZ}(G)) \\ &= D_{(n+1),Z/pZ}(G) . \end{aligned}$$

3. APPLICATIONS

We can apply the results of Section 1 to calculate integral dimension subgroups $D_{n,Z}(G)$ in certain cases.

3.1 Corollary. *If* G *is a group with its lower central factors* $\gamma_i(G)/\gamma_{i+1}(G)$ *torsion-free for all* $i \geq 1$, *then* $D_{i,Z}(G) = \gamma_i(G)$ *for all* $i \geq 1$.

Proof. If $\gamma_i(G)/\gamma_{i+1}(G)$ $(i \geq 1)$ are all torsion-free, then clearly

$$\sqrt{\gamma_i(G)} = \gamma_i(G) \quad \text{for all} \quad i \geq 1 .$$

Hence $D_{i,\mathbb{Q}}(G) = \gamma_i(G)$ (Theorem 1.5). As $\gamma_i(G) \subseteq D_{i,Z}(G) \subseteq D_{i,\mathbb{Q}}(G)$, the result is proved.

Corollary 3.1 applies, in particular, to free groups (see for example [25], Theorem 11.2.4).

We thus have the well known result of Magnus on dimension subgroups of free groups:

3.2 Corollary [48]. If F is a free group, then $D_{i,Z}(F) = \gamma_i(F)$ for all $i \geq 1$.

3.3 Corollary. Let p be a prime. If G is a group with $G^p = 1$, then $D_{i,Z}(G) = \gamma_i(G)$ for all $i \geq 1$.

Proof. If $G^p = 1$, then $G_{i,p} = \gamma_i(G)$ for all $i \geq 1$. Since $\gamma_i(G) \subseteq D_{i,Z}(G) \subseteq D_{i,Z/pZ}(G) = G_{i,p}$ (Theorem 1.9 and Chapter II, Proposition 1.2), the result follows.

3.4 Remarks. (i) We can derive similar conclusions about the integral Lie dimension subgroups $D_{(i),Z}(G)$ from the results of Section 2.

(ii) It is known (Cohn [11], see also Losey [42]) that whenever the lower central series of a group G has the property:

$$x \in \gamma_i(G) \ , \ x \notin \gamma_{i+1}(G) \ , \ x^k \in \gamma_{i+1}(G) \Rightarrow x^k = 1 \ ,$$

then $D_{n,Z}(G) = \gamma_n(G)$ for all $n \geq 1$. This applies, in particular, when the hypothesis of Corollary 3.1 or 3.3 is satisfied.

CHAPTER V

POLYNOMIAL MAPS ON GROUPS AND DIMENSION SUBGROUPS

Polynomial maps on groups are those maps from groups to Abelian groups whose linear extension to the integral group ring vanishes on some power of the augmentation ideal. These maps occur in the literature in several different contexts ([8], [85]). They have proved to be an effective tool for the investigation of dimension subgroups. In this Chapter we study the polynomial maps and calculate some integral and modular dimension subgroups. (For a report on polynomial maps, see also [66].)

1. POLYNOMIAL MAPS

1.1 Definition. Let M be a monoid (an associative multiplicative system with identity) and G an additive Abelian group. A map $f : M \to G$ is called a polynomial map of degree $\leq n$ if the linear extension of f to $Z(M)$, the integral monoid ring of M, vanishes on $\Delta_Z^{n+1}(M)$, where $\Delta_Z(M)$ is the augmentation ideal of $Z(M)$. (Given a monoid M and a commutative ring R with identity, the monoid ring $R(M)$ of M over R is the free R-module on the set M with multiplication defined distributively using the monoid multiplication in M. As for group rings, the augmentation ideal $\Delta_R(M)$ is the kernel of the unit augmentation $\varepsilon : R(M) \to R$.)

1.2 Notation. We denote by $P_n(M,G)$ the set of all polynomial maps $f : M \to G$ of degree $\leq n$ from a monoid M into an Abelian group G.

Let $f_1, f_2 \in P_n(M,G)$. Then the map $f : M \to G$ given by $f(x) = f_1(x) + f_2(x)$, $x \in M$, also lies in $P_n(M,G)$. We can therefore, define addition in the set $P_n(M,G)$ by setting $f_1 + f_2 = f$. With this addition $P_n(M,G)$ is an Abelian group which is isomorphic to $\mathrm{Hom}(Z(M)/\Delta_Z^{n+1}(M), G)$ (see (1.3 (ii)) below).

1.3 Examples

(i) Let M be a monoid and G an Abelian group. Then every (monoid) homomorphism $\alpha : M \to G$ is a polynomial map of degree ≤ 1.

(ii) Let M be a monoid. Then, for every $n \geq 1$, the map $\lambda_n : M \to Z(M)/\Delta_Z^{n+1}(M)$ given by

$$\lambda_n(x) = x + \Delta_Z^{n+1}(M), \quad x \in M,$$

is a polynomial map of degree $\leq n$. This is the <u>universal polynomial map of degree</u> $\leq n$ in the sense that for every polynomial map $f : M \to G$ of degree $\leq n$, there exists a unique homomorphism $\varphi : Z(M)/\Delta_Z^{n+1}(M) \to G$ such that $f = \varphi \circ \lambda_n$. This yields that

$$P_n(M,G) \cong \mathrm{Hom}(Z(M)/\Delta_Z^{n+1}(M), G) .$$

(iii) Let $T : A \to A$ be a functor of degree $\leq n$ in the sense of Eilenberg-Maclane [17] on the category A of Abelian groups. Then for every $M , N \in A$, <u>the map</u>

$$T_{MN} : \mathrm{Hom}(M,N) \to \mathrm{Hom}(T(M), T(N))$$

<u>given by</u>

$$T_{MN}(\theta) = T(\theta) , \quad \theta : M \to N ,$$

<u>is a polynomial map of degree</u> $\leq n$.

(iv) The <u>n-th symmetric product</u> $SP^n(M)$ <u>of an Abelian group</u> M is defined as follows:

Let F be the free Abelian group generated by the symbols $u(m_1,m_2,\dots,m_n)$, $m_i \in M$, $i = 1,2,\dots,n$. Let R be the subgroup generated by the elements of the type

(a) $\quad u(m_1,\dots,m_{i-1},m_i m_{i+1},m_{i+2},\dots,m_{n+1})$

$\quad - u(m_1,\dots,m_{i-1},m_i,m_{i+2},\dots,m_{n+1})$

$\quad - u(m_1,\dots,m_{i-1},m_{i+1},m_{i+2},\dots,m_{n+1})$, $i = 1,2,\dots,n$.

(b) $\quad u(m_1,m_2,\dots,m_n) - u(m_{\pi(1)},m_{\pi(2)},\dots,m_{\pi(n)})$,

Where π is a permutation of the integers $1,2,\dots,n$.

The n-th symmetric product $SP^n(M)$ is defined to be the quotient F/R. We write $m_1 \hat{\otimes} m_2 \hat{\otimes} \dots \hat{\otimes} m_n$ for the coset of $u(m_1,m_2,\dots,m_n)$. Note that a general element of $SP^n(M)$ is a finite sum of the form

$$\Sigma\, m_1 \hat{\otimes} m_2 \hat{\otimes} \dots \hat{\otimes} m_n .$$

Vermani [98] has shown the following:

<u>For each</u> $n \geq 1$, <u>the map</u> $\varphi_n : M \to SP^n(M)$ <u>given by</u>

$$\varphi_n(x) = \underbrace{x \hat{\otimes} x \hat{\otimes} \dots \hat{\otimes} x}_{n \text{ factors}} , \quad x \in M ,$$

<u>is a polynomial map of degree</u> $\leq n$.

<u>Proof</u> (Sandling). Extend φ_n to $Z(M)$ by linearity. If

$x_1, x_2, \dots, x_k \in M$, then $\varphi_n((1-x_1)(1-x_2)\dots(1-x_k)) = \varphi_n(\Sigma(-1)^S X_S)$ where the sum is taken over all subsets S of $\{1,2,\dots,k\}$ with $(-1)^S$ defined as $(-1)^s$, $s = |S|$, and X_S defined as the product of the elements x_i , $i \in S$, taken in the natural order. Expanding by linearity we obtain

$$\Sigma(-1)^S X_S^{\hat{\otimes} n}.$$

The n-fold symmetric power of X_S may be expanded as $\Sigma\binom{n}{\pi}W(\pi)$, where the sum is over all partitions $\pi = \{j_1, \dots, j_k\}$ of n with $j_i = 0$ if $i \notin S$, $W(\pi)$ is the product

$$x_1 \hat{\otimes} \dots \hat{\otimes} x_1 \hat{\otimes} \dots \hat{\otimes} x_k \hat{\otimes} \dots \hat{\otimes} x_k$$

in which x_i appears j_i times, and where $\binom{n}{\pi}$ counts the number of times the term $W(\pi)$ appears in the expansion.

Using these expressions for each S , we obtain

$$\Sigma(-1)^S \Sigma\binom{n}{\pi}W(\pi).$$

Reversing the order of summation, this becomes

$$\Sigma\binom{n}{\pi}W(\pi) \quad \Sigma \quad (-1)^S$$

where now the first summation is taken over all partitions of n and the second over all subsets S containing $S(\pi) = \{i | j_i \neq 0\}$. Since $|S(\pi)| \leq n$ for all π , the sums $\Sigma(-1)^S$ are 0 if $k > n$. In particular, for $k = n + 1$, the image is 0 as required.

Note also that, for $k = n$, the only term not having 0 as coefficient is that corresponding to the partition $j_i = 1$, $i = 1, \dots, n$. In this case, $\binom{n}{\pi}$ which is in general given by

$$\binom{n}{\pi} = \binom{n}{j_1}\binom{n-j_1}{j_2}\dots\binom{n-j_1-j_2-\dots-j_{k-1}}{j_k}$$

is equal to $n!$ so that we obtain

$$\varphi_n((1-x_1)\dots(1-x_n)) = (-1)^n n!(x_1 \hat{\otimes} \dots \hat{\otimes} x_n).$$

(v) Let $M = \{1, t, t^2, \dots\}$ be the infinite cyclic monoid generated by t. Then the monoid ring $R(M)$ of M over any commutative ring R with identity consists of 'polynomials' $\sum_{r_i \in R} r_i t^i$ and the augmentation ideal $\Delta_R(M)$ is the ideal generated by the element $t - 1$. Let $f(X) = a_o + a_1 X + \dots + a_n X^n \in R[X]$ be a polynomial of degree $\leq n$. Then <u>the map</u> $\theta : M \to R$ <u>given by</u> $\theta(t^s) = f(s)$, $0 \leq s \in Z$, <u>is</u>

<u>a polynomial map of degree</u> $\leq n$.

<u>Proof</u>. Induct on $n \geq 0$. If $n = 0$, then $f(X) = a_0 \in R$. Thus $\theta(t^s) = a_0$ for all $s \geq 0$. Hence θ vanishes on $\Delta_R(M)$. Let $n \geq 1$ and suppose that the statement holds for all polynomials $h(X) \in R[X]$ of degree $n - 1$. Now the polynomial $h(X) = f(X+1)-f(X)$ has degree $\leq n - 1$. Hence, by induction, the map $\varphi : M \to R$ given by $\varphi(t^s) = h(s)$ is a polynomial map of degree $\leq n - 1$.

But
$$\begin{aligned}\varphi((t-1)^s) &= \varphi(\sum_{i=0}^{s} (-1)^i \binom{s}{i} t^{s-i}) \\ &= \sum_{i=0}^{s} (-1)^i \binom{s}{i} \varphi(t^{s-i}) \\ &= \sum_{i=0}^{s} (-1)^i \binom{s}{i} (f(s-i+1) - f(s-i)) \\ &= \sum_{i=0}^{s+1} (-1)^i \binom{s+1}{i} f(s+1-i) = \theta((t-1)^{s+1})\end{aligned}$$

Hence θ vanishes on $\Delta_Z^{n+1}(M)$.

(vi) An <u>integer-valued polynomial</u> $p(X) \in \mathbb{Q}[X]$ is a polynomial with rational coefficients which takes integer values when integers are substituted for the indeterminate X. It is well-known that every integer-valued polynomial $p(X)$ can be written as $p(X) = \Sigma\, a_i \binom{X}{i}$,

where $a_i \in Z$ and $\binom{X}{i} = \dfrac{X(X-1)(X-2)\ldots(X-i+1)}{1.2.3\ldots i}$.

[Integer-valued polynomials in several variables are defined in a similar way].

This can be seen by induction on the degree. The statement clearly holds for polynomials of degree 0. Let $p(X) = \sum_{i=0}^{n} b_i X^i$ be an integer valued polynomial of degree $n > 0$ and suppose the assertion holds for polynomials of smaller degree. The polynomial $q(X) = p(X+1)-p(X)$ is evidently an integer-valued polynomial of degree $\leq n-1$. Hence, by induction hypothesis, $q(X) = \sum_{i=0}^{n-1} a_i \binom{X}{i}$, $a_i \in Z$. Comparing the coefficients of X^{n-1} on both sides, we have

$$nb_n = \frac{a_{n-1}}{(n-1)!}$$

Consider the polynomial $p(X) - a_{n-1}\binom{X}{n}$. This is a polynomial of degree $\leq n-1$. Hence, again by induction ,

$$p(X) - a_{n-1}\binom{X}{n} = \sum_{i=0}^{n-1} c_i \binom{X}{i}, \quad c_i \in Z$$

This shows that $p(X)$ is a polynomial of the required form.

Let $p(X) = \sum_{i=0} a_i \binom{X}{i}$ be an integer-valued polynomial of degree $\leq n$. Since the elements $(X-1)^i$, $i = 0,1,\ldots,$ form a Z-basis of $Z[X]$, we can define a homomorphism $f : Z[X] \to Z$ by setting

$$f((X-1)^i) = \begin{cases} a_i, & i=0,1,\ldots,n \\ 0, & i \geq n+1 \end{cases}$$

and by linearity. Let M be the infinite cyclic monoid $\{1,X,X^2,\ldots\}$ Consider the map $\alpha : M \to Z$ given by $\alpha(X^r) = f(X^r)$. Extending α by linearity to $Z(M)$, we clearly have $\alpha((X-1)^i) = 0$ for $i \geq n+1$. Thus α <u>is a polynomial map of degree</u> $\leq n$ (see Section 2).

(vii) Let R be a commutative ring with identity. Let M be the multiplicative group of lower triangular $n\times n$ matrices

$$\begin{pmatrix} 1 & & & & \\ & 1 & & 0 & \\ & & \cdot & & \\ & * & & \cdot & \\ & & & & \cdot \\ & & & & & 1 \end{pmatrix}$$

over R. Consider the map $\theta : M \to \mathfrak{M}_n(R)$, where $\mathfrak{M}_n(R)$ is the ring of $n\times n$ matrices over R. From matrix multiplication it is clear that if $Y_1,Y_2,\ldots,Y_n \in M$, then

$$\theta((Y_1 - I)(Y_2 - I)\ldots(Y_n - I)) = 0,$$

where I is the identity matrix. Thus θ <u>is a polynomial map of degree</u> $\leq n-1$.

(viii) Let L be a Lie algebra over the rationals and suppose that L is nilpotent of class 2, i.e. $[[x,y],z] = 0$ for all $x,y,z \in L$. For $x,y \in L$ define $x\circ y$ by

$$x\circ y = x + y + \tfrac{1}{2}[x,y].$$

Clearly $\circ$ is an associative operation on L and the zero element of L acts as the multiplicative identity. Let $M = (L,\circ)$ be the monoid with L as the underlying set and $\circ$ as the multiplication. Consider the map $\mu : M \to L$, $\mu(x) = x$, $x \in M$. Extend μ to $Z(M)$ by linearity.

Let 1_M be the multiplicative identity of M . Then

$$\mu((x_1-1_M)(x_2-1_M)(x_3-1_M)) = \mu((x_1\circ x_2\circ x_3-x_1\circ x_2-x_1\circ x_3-x_2\circ x_3+x_1+x_2+x_3-1_M))$$

$$= x_1\circ x_2\circ x_3 - x_1\circ x_2-x_1\circ x_3-x_2\circ x_3+x_1+x_2+x_3$$

$$= \{x_1+x_2+x_3+ \tfrac{1}{2}[x_2,x_3] + \tfrac{1}{2}[x_1,x_2] + \tfrac{1}{2}[x_1,x_3]\}$$

$$- (x_1+x_2+ \tfrac{1}{2}[x_1,x_2]) - (x_1+x_3+ \tfrac{1}{2}[x_1,x_3])$$

$$- (x_2+x_3+ \tfrac{1}{2}[x_2,x_3]) + x_1+x_2+x_3$$

$$= 0$$

Thus μ <u>is a polynomial map of degree</u> ≤ 2 .

2. A CHARACTERIZATION OF POLYNOMIAL MAPS ON GROUPS

2.1 <u>Theorem</u> [8]. <u>Let</u> A <u>be an Abelian group written additively and</u> B <u>an arbitrary group written multiplicatively</u>. <u>Then a map</u> $f : B \to A$ <u>is a polynomial map of degree</u> $\leq r$ <u>if and only if for every choice of</u> $b_1,b_2,\ldots,b_k \in B$, <u>there exist integer-valued polynomials</u> $p_1,p_2,\ldots,p_t$ <u>all of degree</u> $\leq r$ <u>and</u> $a_1,a_2,\ldots,a_t \in A$ <u>such that</u>

$$f(b_1^{m_1}b_2^{m_2}\ldots b_k^{m_k}) = \sum_{i=1}^{t} p_i(m_1,m_2,\ldots,m_k)a_i$$

<u>for all integers</u> $m_1,m_2,\ldots,m_k$.

<u>Proof</u>. Let $f : B \to A$ be a polynomial map of degree $\leq r$. Let $b_1,b_2,\ldots,b_k \in B$. Define elements $a_{i_1i_2\ldots i_k} \in A$ by setting

$$f((1-b_1^{-1})^{i_1}(1-b_2^{-1})^{i_2}\ldots(1-b_k^{-1})^{i_k}) = a_{i_1i_2\ldots i_k}$$

for every k-tuple $(i_1,i_2,\ldots,i_k)$ of non-negative integers with $i_1+ i_2+\ldots i_k \leq r$. We note that in $Z(B)$

$$b^m = \sum_{i=0}^{r}\binom{m+i-1}{i}(1-b^{-1})^i \mod \Delta_Z^{r+1}(B)$$

for every $b \in B$ and $m \in Z$. Thus if $m_1, m_2, \ldots, m_k \in Z$, then

$$b_1^{m_1} b_2^{m_2} \ldots b_k^{m_k} = \sum_{0 \leq i_1 + \ldots + i_k \leq r} \binom{m_1+i_1-1}{i_1} \ldots \binom{m_k+i_k-1}{i_k} \left(1-b_1^{-1}\right)^{i_1} \ldots$$

$$\ldots (1-b_k^{-1})^{i_k} \bmod \Delta_Z^{r+1}(B) .$$

Hence

$$f\left(b_1^{m_1} b_2^{m_2} \ldots b_k^{m_k}\right) = \Sigma\, p_{i_1 \ldots i_k}(m_1, m_2, \ldots, m_k)\, a_{i_1 \ldots i_k}$$

where $p_{i_1 \ldots i_k}(X_1, X_2, \ldots, X_k) = \binom{X_1+i_1-1}{i_1} \ldots \binom{X_k+i_k-1}{i_k}$

which is clearly an integer-valued polynomial.

To prove the converse we proceed by induction on r. If $r = 0$, then f is a constant map and so it is trivially a polynomial map of degree 0. Suppose the result holds for $r-1$ $(r \geq 1)$. Let $b \in B$. Consider the map $g : B \to A$ given by $g(x) = f(xb) - f(x)$. Let $b_1, b_2, \ldots, b_k \in B$. Then, by hypothesis, there exist integer-valued polynomials $p_i(X_1, X_2, \ldots, X_{k+1})$ and elements $a_i \in A$ such that

$$f(b_1^{m_1} b_2^{m_2} \ldots b_k^{m_k} b^{m_{k+1}}) = \sum_{i=1}^{t} p_i) m_1, m_2, \ldots, m_{k+1}) a_i$$

for all $m_1, m_2, \ldots, m_{k+1} \in Z$. Thus

$$g(b_1^{m_1} b_2^{m_2} \ldots b_k^{m_k}) = f(b_1^{m_1} \ldots b_k^{m_k} b) - f(b_1^{m_1} \ldots b_k^{m_k})$$

$$= \sum_{i=1}^{t} \{p_i(m_1, \ldots, m_k, 1) - p_i(m_1, \ldots, m_k, 0)\} a_i .$$

Since the polynomials

$$q_i(X_1,\dots,X_k) = p_i(X_1,\dots,X_k,1) - p_i(X_1,\dots,X_k,0)$$

are all integer-valued of degree $\leq r-1$, induction gives that g vanishes on $\Delta_Z^r(B)$. As b was arbitrary, we conclude that f vanishes on $\Delta_Z^{r+1}(B)$, i.e. f is a polynomial map of degree $\leq r$.

3. WREATH PRODUCT

3.1 Standard restricted wreath product. Let A and B be two groups, A^B the set of maps $f : B \to A$. With componentwise multiplication

$$(fg)(b) = f(b)g(b)\ ,\ b \in B\ ,\ f,\ g \in A^B\ ,$$

A^B is a group. We define the support $\sigma(f)$ of f by

$$\sigma(f) = \{b \in B \mid f(b) \neq 1\}\ .$$

Let $A^{(B)} = \{f \in A^B \mid \sigma(f)$ is finite$\}$. Then $A^{(B)}$ forms a subgroup of A^B . The standard restricted wreath product $A \text{ wr } B$ of A by B is the split extension of $A^{(B)}$ by B , where the action of B on $A^{(B)}$ is given by

$$f^b(b') = f(b'b^{-1})\ ,\ f \in A^{(B)}\ ,\ b,\ b' \in B\ .$$

3.2 α-Central Series. Let $W = A \text{ wr } B$ and $\{\zeta_i(W)\}_{i\geq 0}$ be the upper central series of W . The α-central series of W is the series $\{\alpha_n(W)\}_{n\geq 1}$ defined by

$$\alpha_i(W) = \zeta_i(W) \cap A^{(B)}\ .$$

α-central series have been investigated by Meldrum ([51], [52], [53]) and Buckley [8]. This type of relation between wreath products and group rings has also been studied by Sandling ([81], [82]) and Shield [87].

The following result relates the notions of α-central series and polynomial maps.

3.3 Theorem [8]. *If A and B are two groups such that A is Abelian and B is finite, then*

$$\alpha_{n+1}(A \text{ wr } B) = P_n(B,A) = \text{Hom}(Z(B)/\Delta_Z^{n+1}(B),A)\ .$$

Proof. Let $f : B \to A$ be a map in $A^{(B)}$. If $b \in B$, then the commutator $(f,b) = f^{-1}b^{-1}fb \in A^{(B)}$ is the map which sends an element $x \in B$ into the element $x(b^{-1}-1)$. More generally, induction shows that the commutator $(\ldots((f,b_1),b_2),\ldots,b_{n+1})$ is the map which sends x into $x(b_{n+1}^{-1}-1)(b_n^{-1}-1)\ldots(b_1^{-1}-1)$. It is easy to check that a map $f : B \to A$ belongs to $\alpha_{n+1}(A \text{ wr } B)$ if and only if

$$(\ldots((f,b_1),b_2),\ldots,b_{n+1}) = 1$$

for all $b_1,b_2,\ldots,b_{n+1} \in B$. Thus $f \in \alpha_{n+1}(A \text{ wr } B)$ if and only if

$$f(x(b_{n+1}^{-1}-1)(b_n^{-1}-1)\ldots(b_1^{-1}-1)) = 0$$

for all $x, b_1,b_2,\ldots,b_{n+1} \in B$, i.e. if and only if $f \in P_n(B,A)$. For the second isomorphism see 1.3 (ii).

4. RELATIONSHIP WITH DIMENSION SUBGROUPS

Let T denote the additive (divisible) Abelian group $\mathbb{Q}/\mathbb{Z}$ of rationals mod 1. Let $D_n(G)$ denote the n-th dimension subgroup of G over integers, i.e. $D_{n,\mathbb{Z}}(G)$. We maintain these notations for the rest of this Chapter. Note that T has the property that, for Abelian groups, $\mathrm{Hom}(G,T) = 0$ if and only if $G = 0$.

4.1 Proposition. *For every group* G *and integer* $n \geq 0$,

$$D_{n+1}(G) = \{x \in G \mid \varphi(x) = \varphi(1), \text{ for all } \varphi \in P_n(G,T)\}.$$

Proof. Let $\delta_{n+1}(G) = \{x \in G \mid \varphi(x) = \varphi(1) \text{ for all } \varphi \in P_n(G,T)\}$. If $x \in D_{n+1}(G)$, then $x-1 \in \Delta_{\mathbb{Z}}^{n+1}(G)$ and therefore $\varphi(x-1) = 0$ for all $\varphi \in P_n(G,T)$. This shows that $\delta_{n+1}(G) \supseteq D_{n+1}(G)$. Let $x \in \delta_{n+1}(G)$. If $x \notin D_{n+1}(G)$, then $x-1+\Delta_{\mathbb{Z}}^{n+1}(G)$ is a nonzero element of $\mathbb{Z}(G)/\Delta_{\mathbb{Z}}^{n+1}(G)$. Therefore, we can find a homomorphism $\varphi : \mathbb{Z}(G)/\Delta_{\mathbb{Z}}^{n+1}(G) \to T$ such that $\varphi(x-1+\Delta_{\mathbb{Z}}^{n+1}(G)) \neq 0$. Consider the map $\eta : G \to T$ given by $\eta(y) = \varphi(y-1+\Delta_{\mathbb{Z}}^{n+1}(G))$, $y \in G$. Then η is evidently a polynomial map of degree $\leq n$. However, $\eta(x) \neq \eta(1)$. This contradiction proves that $D_{n+1}(G) \supseteq \delta_{n+1}(G)$. Hence $D_{n+1}(G) = \delta_{n+1}(G)$.

In view of Proposition 4.1 polynomial maps have proved an effective tool for the calculation of integral dimension subgroups.

Let G be a nilpotent group of class n and let $G = \gamma_1(G) \supseteq \gamma_2(G) \supseteq \ldots \supseteq \gamma_n(G) \supseteq \gamma_{n+1}(G) = 1$ be its lower central series. Suppose that $\gamma_n(G) \cap D_{n+1}(G) = 1$. Consider the map

$$\theta_n : \gamma_n(G) \to \mathbb{Z}(G)/\Delta_{\mathbb{Z}}^{n+1}(G)$$

given by $\theta_n(x) = x-1+\Delta_Z^{n+1}(G)$. Clearly θ_n is a homomorphism and because we are assuming that $\gamma_n(G) \cap D_{n+1}(G) = 1$, θ_n is one-one. Hence for every homomorphism $\alpha : \gamma_n(G) \to T$, there exists a homomorphism $\varphi : Z(G)/\Delta_Z^{n+1}(G) \to T$ such that $\alpha = \varphi\circ\theta_n$. Define $\eta : G \to T$ by $\eta(x) = \varphi(x-1+\Delta_Z^{n+1}(G))$. Then η is a polynomial map of degree $\leqslant n$ and its restriction to $\gamma_n(G)$ coincides with α . Thus we have proved:

4.2 Proposition. *If G is a nilpotent group of class n satisfying $\gamma_n(G) \cap D_{n+1}(G) = 1$, then every homomorphism $\alpha : \gamma_n(G) \to T$ can be extended to a polynomial map $\eta : G \to T$ of degree $\leqslant n$.*

The converse of 4.2 also holds. If every homomorphism $\alpha : \gamma_n(G) \to T$ can be extended to a polynomial map $\eta : G \to T$ of degree $\leqslant n$, then $\gamma_n(G) \cap D_{n+1}(G) = 1$. For, suppose $x \in \gamma_n(G) \cap D_{n+1}(G)$ and $x \neq 1$. Then we can find a homomorphism $\alpha : \gamma_n(G) \to T$ such that $\alpha(x) \neq 0$. Extend α to a polynomial map $\eta : G \to T$ of degree $\leqslant n$. Now $\eta(x) = \alpha(x) \neq 0$, while $\eta(x) = \eta(1)$ (Proposition 4.1) and $\eta(1) = \alpha(1) = 0$. Hence no non-identity element can lie in $\gamma_n(G) \cap D_{n+1}(G)$ and we have

4.3 Proposition. *If G is a group of class n such that every homomorphism $\alpha : \gamma_n(G) \to T$ can be extended to a polynomial map $\eta : G \to T$ of degree $\leqslant n$, then $\gamma_n(G) \cap D_{n+1}(G) = 1$.*

Before taking up the computation of dimension subgroups in the next section, we illustrate a useful reduction (due to G. Higman) which will be employed repeatedly.

4.4 Theorem. *If G is a group such that $D_n(G) \neq \gamma_n(G)$, then there is a subquotient N of G with the properties*

(a) *N is a finite p-group of class $\leqslant n-1$;*

(b) $D_n(N) \neq 1$.

Proof. Let G be a group having an element x such that $x-1 \in \Delta_Z^n(G)$, $x \notin \gamma_n(G)$. An expression for $x-1$ as an element of $\Delta_Z^n(G)$ involves only a finite number of group elements. Let H be the subgroup of G generated by these elements. Then H is a finitely generated group, $x \in D_n(H)$, $x \notin \gamma_n(H)$. Let $K = H/\gamma_n(H)$. As dimension subgroups are preserved under homomorphisms, $\bar{x} = x\,\gamma_n(H) \in D_n(K)$ and $\bar{x} \notin \gamma_n(K) = 1$. Now K is a finitely generated nilpotent group. It is a well-known result of Gruenberg [19] that a finitely generated nilpotent group is residually a prime power group (see Chapter VI, 2.10). Hence there exists a normal subgroup L, say, such that $\bar{x} \notin L$ and $K/L = N$ is a prime power group. Now $\bar{x}L \in D_n(N)$ and $\bar{x}L \notin \gamma_n(N) = 1$. As N by construction is a quotient of a subgroup of G, the Theorem is proved.

5. INTEGRAL DIMENSION SUBGROUPS

Proposition 4.2 and 4.3 show that if we compute the integral dimension subgroups by induction on the class, then we are faced with the problem of extending homomorphisms from the last non-identity term in the lower central series to polynomial maps on the whole group. We begin with an investigation of this extension problem.

Let $1 \to N \xrightarrow{i} \Pi \xrightarrow{\beta} G \to 1$ be an exact sequence of groups with N Abelian and i the inclusion map. Let M be an Abelian group. We use the notation $(N,n\Pi) = (\ldots((\underbrace{N,\Pi),\Pi),\ldots,\Pi)}_{n \text{ terms}}$.

5.1 Theorem [67]. A homomorphism $\alpha : N \to M$ can be extended to a map $\varphi : \Pi \to M$ whose linear extension to $Z(\Pi)$ vanishes on $\Delta_Z^{n+1}(\Pi) + \Delta_Z(\Pi)\Delta_Z(N)$ if and only if

(i) there exists a transversal $\{w(g)\}_{g\in G}$, $w(1) = 1$, for G in Π and a map $\chi : G \to M$, $\chi(1) = 0$, such that

$$\alpha(W(g_1,(g_2-1)(g_3-1)\ldots(g_{n+1}-1))) = \chi((g_1-1)(g_2-1)\ldots(g_{n+1}-1)),$$

$g_i \in G$, $i = 1,2,\ldots,n+1$, where $W(g_1,g_2) = w(g_1,g_2)^{-1}w(g_1)w(g_2)$: $G\times G \to N$ is the 2-cocycle determined by the transversal w, N is regarded as a right G-module via conjugation in Π and both W and χ have been extended to $Z(G)\times Z(G)$ and $Z(G)$ respectively by linearity;

(ii) α vanishes on $(N,n\Pi)$.

Proof. Let α be a homomorphism and suppose both (i) and (ii) are satisfied. Every element of Π can be written uniquely as $w(g)z$, $g \in G$, $z \in N$. Define $\varphi : \Pi \to M$ by setting $\varphi(w(g)z) = \alpha(z) - \chi(g)$, $g \in G$, $z \in N$. We extend φ to $Z(\Pi)$ by linearity. Then

$$\varphi((w(g_1)z_1-1)(w(g_2)z_2-1)) = \alpha(W(g_1,g_2)) + \alpha(z_1(g_2-1)) \\ -\chi((g_1-1)(g_2-1)),$$

$g_1,g_2 \in G$, $z_1,z_2 \in N$. Induction gives

$$\varphi((w(g_1)z_1-1)(w(g_2)z_2-1)\ldots(w(g_{m+1})z_{m+1}-1))$$

$= \alpha(W(g_1, (g_2-1)(g_3-1)\dots(g_{m+1}-1))) + \alpha(z_1\cdot(g_2-1)(g_3-1)\dots(g_{m+1}-1))$

$-\chi((g_1-1)(g_2-1)\dots(g_{m+1}-1)), g_i \in G,\ z_i \in N,\ i = 1,2,\dots,m+1$

for all $m \geq 1$.

Now we note that

$$z_1\cdot(g_2-1)\dots(g_{m+1}-1) \in (N, m\Pi) .$$

Hence we have

$$\varphi((w(g_1)z_1-1)(w(g_2)z_2-1)\dots(w(g_{n+1})z_{n+1}-1)) = 0 .$$

Thus φ vanishes on $\Delta_Z^{n+1}(\Pi)$. Also, for $g \in G$, $x,y \in N$, we have

$$\varphi((w(g)x-1)(y-1)) = 0 .$$

Thus φ vanishes on $\Delta_Z(\Pi)\Delta_Z(N)$.

Conversely, suppose α can be extended to a map $\varphi : \Pi \to M$ which vanishes on $\Delta_Z^{n+1}(\Pi) + \Delta_Z(\Pi)\Delta_Z(N)$ when extended by linearity to $Z(\Pi)$. As $(N, n\Pi) \subseteq 1 + \Delta_Z^{n+1}(\Pi)$, α must vanish on $(N, n\Pi)$. Thus $\alpha(z\cdot(g_1-1)(g_2-1)\dots(g_n-1)) = 0$ for all $z \in N$, $g_1, g_2, \dots, g_n \in G$. Let $\{w(g)\}_{g\in G}$ be any transversal for G in Π and $W : G \times G \to N$ be the corresponding 2-cocycle. Then $\varphi((w(g)-1)(z-1)) = 0$ for all $g \in G$ and $z \in N$, since φ vanishes on $\Delta_Z(\Pi)\Delta_Z(N)$.

Thus $\varphi(w(g)z) = \varphi(w(g)) + \varphi(z)$, $g \in G$, $z \in N$

$= -\chi(g) + \alpha(z)$, where $\chi(g) = -\varphi(w(g))$.

Let $g_i \in G$, $z_i \in N$, $i = 1,2,\dots,n+1$. Then we have

$0 = \varphi((w(g_1)z_1-1)(w(g_2)z_2-1)\dots(w(g_{n+1})z_{n+1}-1))$

$= \alpha(W(g_1, (g_2-1)(g_3-1)\dots(g_{n+1}-1))) + \alpha(z_1\cdot(g_2-1)(g_3-1)\dots(g_{n+1}-1))$

$-\chi((g_1-1)(g_2-1)\dots(g_{n+1}-1))$

$= \alpha(W(g_1, (g_2-1)(g_3-1)\dots(g_{n+1}-1))) - \chi((g_1-1)(g_2-1)\dots(g_{n+1}-1))$.

This completes the proof of the Theorem.

5.2 Definition. We say that a group Π has the dimension property if its lower central series $\Pi = \gamma_1(\Pi) \supseteq \gamma_2(\Pi) \supseteq \dots \supseteq \gamma_n(\Pi) \supseteq \dots$ coincides with its integral dimension series $\Pi = D_1(\Pi) \supseteq D_2(\Pi) \supseteq \dots \supseteq D_n(\Pi) \supseteq \dots$ i.e. $\gamma_n(\Pi) = D_n(\Pi)$ for all $n \geq 1$. For example, free groups have the dimension property (see Chapter IV, Section 3).

By Theorem 4.4, if G is a group such that every prime power subquotient of G has the dimension property, then so has G itself.

It is well-known that the Abelian groups have the dimension property (see Chapter II, Example 1.5). In order to illustrate the present theory, we prove the following

5.3 Proposition. For every group G, $D_2(G) = \gamma_2(G)$.

Proof. It clearly suffices to prove that for an Abelian group G, $D_2(G) = 1$. Suppose $1 \neq x \in D_2(G)$. Let $N = \langle x \rangle$ be the subgroup generated by x. Then we can find a homomorphism $\alpha : N \to T$ such that $\alpha(x) \neq 0$. Now T is a divisible Abelian group. Therefore, α can be extended to a homomorphism $\varphi : G \to T$. It is easily seen that the linear extension of φ to $Z(G)$ vanishes on $\Delta_Z^2(G)$. Consequently, $\varphi(x-1) = 0$. But $\varphi(x-1) = \varphi(x) - \varphi(1) = \alpha(x) \neq 0$. This is a contradiction. Hence $D_2(G) = 1$.

It has been shown by Sandling [79] that Abelian-by-cyclic groups have the dimension property. The following Theorem improves this result.

5.4 Theorem ([67], [79]). If N is a normal Abelian subgroup of a group Π with Π/N cyclic, then $\gamma_n(\Pi) = \Pi \cap (1+\Delta_Z^n(\Pi) + \Delta_Z(\Pi)\Delta_Z(N))$ for all $n \geq 1$.

Proof. The result is trivial for $n = 1$, and it holds for $n = 2$ by Proposition 5.3. Let $n > 2$. By factoring with $\gamma_n(\Pi)$, if necessary, we can assume that $\gamma_n(\Pi) = 1$. Let $1 \neq x \in \Pi \cap (1+\Delta_Z^n(\Pi) + \Delta_Z(\Pi)\Delta_Z(N))$. Then $x \in \gamma_2(\Pi) \subseteq N$. Thus we can find a homomorphism $\alpha : N \to T$ such that $\alpha(x) \neq 0$. Let $G = \Pi/N$ and λ be a generator of G. Choose $w(\lambda)$, a representative for λ in Π, arbitrarily and set $w(\lambda^i) = w(\lambda)^i$ for all i if λ is of infinite order and for $0 \leq i <$ order of λ otherwise. Then $W : G \times G \to N$, the 2-cocycle determined by this choice of the transversal $\{w(g)\}_{g \in G}$ has the property that

$$W(g_1,g_2)\cdot g_3 = W(g_1,g_2), g_1,g_2,g_3 \in G.$$

Then $\alpha(W(g_1,g_2)) : G \times G \to T$ is a 2-cocycle with T regarded as a trivial G-module. As the second cohomology group $H^2(G,T) = 0$ ([30], Proposition 7.1), there exists a map $\chi : G \to T$ such that

$$\alpha(W(g_1,g_2)) = -\chi(g_2) + \chi(g_1 g_2) - \chi(g_1), \quad g_1,g_2 \in G.$$

This gives

$$\alpha(W(g_1,(g_2-1)(g_3-1)\dots(g_{n+1}-1))) = \chi((g_1-1)(g_2-1)\dots(g_{n+1}-1)),$$

$g_i \in G$, $i = 1,2,\dots,n+1$. As $(N,(n-1)\Pi) \subseteq \gamma_n(\Pi) = 1$, we conclude that both conditions of Theorem 5.1 are satisfied. Hence α can be extended to a map $\varphi : \Pi \to T$ which vanishes on $\Delta_Z^n(\Pi) + \Delta_Z^n(\Pi)\Delta_Z(N)$. But then $\alpha(x) = \varphi(x) = \varphi(x-1) = 0$ which is a contradiction. Hence $\gamma_n(\Pi) = 1$ implies that

$$\Pi \cap (1+\Delta_Z^n(\Pi) + \Delta_Z(\Pi)\Delta_Z(N)) = 1$$

and the Theorem is proved.

Another case in which Theorem 5.1 can be coveniently applied is that of a split extension.

5.5 <u>Theorem</u> ([67], [79]). <u>If</u> Π <u>is a split extension of its normal Abelian subgroup</u> N <u>by a group</u> G, <u>then</u>

$$\Pi \cap (1+\Delta_Z^n(\Pi) + \Delta_Z(\Pi)\Delta_Z(N)) = D_n(G)(N,(n-1)\Pi)$$

<u>for all</u> $n \geqslant 1$.

<u>Proof</u>. That the right hand side is contained in the left hand side is obvious. Let $n \geqslant 2$ and $\pi \in \Pi \cap (1+\Delta_Z^n(\Pi)+\Delta_Z(\Pi)\Delta_Z(N))$ and suppose that $\pi = gz$, $g \in G$, $z \in N$. By projecting to G it is immediate that $g \in D_n(G)$. Thus $z \in 1+\Delta_Z^n(\Pi)+\Delta_Z(\Pi)\Delta_Z(N)$ and we have to prove that $z \in (N,(n-1)\Pi)$. Suppose $z \notin (N,(n-1)\Pi)$. Then we can find a homomorphism $\alpha : N \to T$ for which $\alpha(z) \neq 0$ while restriction of α to $(N,(n-1)\Pi)$ is 0. Thus α satisfies condition (ii) of Theorem 5.1. Further, since Π is a split extension of N by G, we can choose $\{w(g)\}_{g\in G}$ so that the corresponding 2-cocycle $W(g_1,g_2) = 0$. Thus, taking $\chi(g) = 0$ for all $g \in G$ the condition (i) is satisfied trivially. Hence α can be extended to a map $\varphi : \Pi \to T$ whose linear extension to $Z(\Pi)$ vanishes on $\Delta_Z^n(\Pi)+\Delta_Z(\Pi)\Delta_Z(N)$. But $\alpha(z) = \varphi(z-1)=0$, a contradiction. Hence $z \in (N,(n-1)\Pi)$ and the proof is complete.

If $1 \to N \xrightarrow{i} \Pi \xrightarrow{\beta} G \to 1$ is a central extension, then 5.1(ii) is trivially satisfied. In this case 5.1(i) can be best expressed in terms of the notion of <u>polynomial 2-cocycles</u> which we now introduce.

5.6 <u>Definition</u>. Let G be a group, M a trivial G-module. Then a map $f : G \times G \to M$ is called a (normalized) <u>polynomial 2-cocycle of degree</u> $\leqslant n$ if

(i) $f(x,y) = 0$ whenever x or y is 1;

(ii) $f(y,z) - f(xy,z) + f(x,yz) - f(x,y) = 0$ for all $x,y,z \in G$;

(iii) for every $x \in G$, the map $f_x : G \to M$ defined by $f_x(y) = f(x,y)$ is a polynomial map of degree $\leq n$.

Let $f : G \times G \to M$ be a polynomial 2-cocycle. We can extend f to $Z(G) \times Z(G)$ by linearity. The condition (ii) can then be expressed as

$$f((x-1)(y-1),z-1) = f(x-1,(y-1)(z-1)).$$

Thus for a polynomial 2-cocycle $f : G \times G \to M$ of degree $\leq n$, the map $f(-,x) : G \to M$ is also a polynomial map of degree $\leq n$ for every $x \in G$.

Let $P_nH^2(G,M)$ be the subset of $H^2(G,M)$ consisting of those elements of $H^2(G,M)$ which possess polynomial 2-cocycles of degree $\leq n$ as representatives. Clearly $P_nH^2(G,M)$ is a subgroup of $H^2(G,M)$ and we have a filtration

$$0 = P_0H^2(G,M) \subseteq P_1H^2(G,M) \subseteq P_2H^2(G,M) \subseteq \ldots \subseteq P_nH^2(G,M) \subseteq \ldots$$

of $H^2(G,M)$. This filtration has been studied by Passi-Stammbach [73] for $M = T$ (the additive group of rationals mod 1) and by Sharma [86] for $M = Z/pZ$ (p prime).

We observe that for a central extension 5.1(i) amounts to saying that the 2-cocycle

$$\gamma(g_1,g_2) = \alpha(W(g_1,g_2)) - \chi((g_1-1)(g_2-1)), \quad g_1,g_2 \in G$$

is a polynomial 2-cocycle of degree $\leq$ n. Equivalently, this means that the cohomology class $\xi \in H^2(G,M)$ induced by the homomorphism $\alpha : N \to M$ should belong to $P_nH^2(G,M)$. This motivates the investigation of conditions under which there exists n such that $P_nH^2(G,M) = H^2(G,M)$, especially when M is T.

5.7 Lemma. If A and B are two groups such that $P_nH^2(A,M)=H^2(A,M)$ and $P_nH^2(B,M) = H^2(B,M)$, then $P_nH^2(A\oplus B,M) = H^2(A\oplus B,M)$.

Proof. Let $\xi \in H^2(A\oplus B,M)$ and $\beta : A \oplus B \times A \oplus B \to M$ be a representative 2-cocycle of ξ. Define $f : A \oplus B \times A \oplus B \to M$ by

$$f(a_1b_1,a_2b_2) = \beta(a_1b_1,a_2b_2) + \chi((a_1b_1-1)(a_2b_2-1))$$

where

$$\chi(ab) = -\beta(b,a), a,a_1,a_2 \in A; \; b,b_1,b_2 \in B.$$

Then f is a 2-cocycle cohomologous to β and has the property that $f(b,a) = 0$ for $b \in B$ and $a \in A$. Using this property of f and the identity

$$f((h_1-1)(h_2-1),h_3) = f(h_1,(h_2-1)(h_3-1)), \text{ for all } h_1,h_2,h_3 \in A \oplus B$$

it is not hard to prove that

(i) $f(a_1b_1,a_2b_2) = f(a_1,a_2) + f(a_1,b_2) + f(b_1,b_2)$, a_1,a_2 A; $b_1,b_2 \in B$

and

(ii) $f(a,b): A \times B \longrightarrow M$ is bilinear.

As $f_{|A\times A}$ and $f_{|B\times B}$ are 2-cocycles, there exist maps $q_1 : A \to M$ and $q_2 : B \to M$ such that

$f(a_1,a_2) + q_1((a_1-1)(a_2-1)): A\times A \to M (a_1,a_2 \in A)$ is a polynomial 2-cocycle of degree $\leq n$ and $f(b_1,b_2) + q_2((b_1-1)(b_2-1)): B\times B \to M$ $(b_1,b_2 \in B)$ is a polynomial 2-cocycle of degree $\leq n$. Let

$$q(ab) = q_1(a) + q_2(b).$$

Then the 2-cocycle

$$f(h_1,h_2) + q((h_1-1)(h_2-1)): A \oplus B \times A \oplus B \to M, (h_1,h_2 \in A \oplus B)$$

which is cohomologous to f, is easily seen to be of degree $\leq n$. Hence $\xi \in P_nH^2(G,M)$ and the Lemma is proved.

5.8 <u>Lemma</u>. *If* G *is a finitely generated Abelian group, then*

$$P_1H^2(G,T) = H^2(G,T).$$

<u>Proof</u>. If G is a cyclic group, then $H^2(G,T) = 0$. Therefore, the result follows by induction on the number of cyclic components of G using Lemma 5.7.

5.9 <u>Theorem</u> ([67], [79]). *For every group* Π,

$$\gamma_3(\Pi) = \Pi \cap (1+\Delta_Z^3(\Pi) + \Delta_Z(\Pi)\,\Delta_Z(\zeta_1(\Pi))),$$

where $\zeta_1(\Pi)$ *denotes the centre of* Π.

<u>Proof</u>. We first observe that the left hand side is contained in the right hand side because $\gamma_3(\Pi) \subseteq D_3(\Pi)$ and that to establish the reverse inclusion it suffices to consider the case when Π is finitely generated with $\gamma_3(\Pi) = 1$. By Theorem 5.1 and Lemma 5.8, it follows that every homomorphism $\alpha : \zeta_1(\Pi) \to T$ can be extended to a map

$\varphi : \Pi \to T$ which vanishes on $\Delta_Z^3(\Pi) + \Delta_Z(\Pi)\Delta_Z(\zeta_1(\Pi))$ when extended by linearity to $Z(\Pi)$. Hence $\zeta_1(\Pi) \cap (1+\Delta_Z^3(\Pi)+\Delta_Z(\Pi)\Delta_Z(\zeta_1(\Pi))) = 1$. Since $\Pi \cap (1+\Delta_Z^3(\Pi) + \Delta_Z(\Pi)\Delta_Z(\zeta_1(\Pi))) \subseteq \gamma_2(\Pi) \subseteq \zeta_1(\Pi)$, the proof is complete.

Theorem 5.9, in particular, gives the third dimension subgroup $D_3(\Pi)$.

5.10 <u>Theorem</u>. <u>For every group</u> Π, $D_3(\Pi) = \gamma_3(\Pi)$.

The result 5.10 is due to G. Higman and D. Rees independently (see also [4], [31], [63], [80]).

Using Theorem 5.10, it is possible to extend Lemma 5.8 to all Abelian groups.

5.11 <u>Theorem</u> [65] <u>For every Abelian group</u> G, $P_1H^2(G,T) = H^2(G,T)$.

<u>Proof</u>. Let $\xi \in H^2(G,T)$. Let $1 \to T \to M \to G \to 1$ be a central extension corresponding to the cohomology class ξ, $T \subseteq M$. As G is Abelian, $\gamma_2(M) \subseteq T$. The exact sequence

$$1 \to T/\gamma_2(M) \to M/\gamma_2(M) \to G \to 1$$

of Abelian groups must split, $T/\gamma_2(M)$ being divisible Abelian. Thus there exists a group N containing $\gamma_2(M)$ in its centre and a commutative diagram

$$\begin{array}{ccccccccc} 1 & \to & \gamma_2(M) & \to & N & \to & G & \to & 1 \\ & & \downarrow i & & \downarrow \beta & & \| & & \\ 1 & \to & T & \to & M & \to & G & \to & 1 \end{array}$$

We assert that $\gamma_2(N) = \gamma_2(M)$. Evidently $\gamma_2(N) \subseteq \gamma_2(M)$. Every element of M can be written as $\beta(n)t$, $n \in N$, $t \in T$. As T is in the centre of M, it follows that $\gamma_2(M) \subseteq \beta(\gamma_2(N)) = \gamma_2(N)$ ($\gamma_2(N) \subseteq \gamma_2(M)$ and the diagram is commutative). Since $D_3(N) = 1$, the homomorphism $i : \gamma_2(N) \to T$ can be extended to a polynomial map of degree ≤ 2 (Proposition 4.2). But this implies (Theorem 5.1) that the element $\xi \in P_1H^2(G,T)$.

6. <u>THE FOURTH INTEGRAL DIMENSION SUBGROUP</u>

If Π is a p-group, $p \neq 2$, then $D_4(\Pi) = \gamma_4(\Pi)$. Moreover, for any group Π, $D_4(\Pi)/\gamma_4(\Pi)$ has exponent at most 2. We present proofs of these results following the methods developed in [63]. For technical details omitted here the reader is referred to [63].

Let G be a finite p-group of class 2. Let $G/\gamma_2(G) = H = C(\lambda_1) \oplus C(\lambda_2) \oplus \ldots \oplus C(\lambda_n)$, where $C(\lambda_i)$ is a cyclic group of order p^{α_i} , say, generated by λ_i , $i = 1,2,\ldots,n$. We choose representatives $\{w(h)\}_{h \in H}$ of elements of H in G as follows:

Choose $w(\lambda_i)$, $i = 1,2,\ldots,n$, without any restriction and set

$$w(\lambda_1^{r_1}\lambda_2^{r_2} \ldots \lambda_n^{r_n}) = w(\lambda_n)^{r_n} w(\lambda_{n-1})^{r_{n-1}} \ldots w(\lambda_1)^{r_1}, \ 0 \leq r_i < p^{\alpha_i} .$$

Let $W(h_1,h_2) = w(h_1)w(h_2)w(h_1h_2)^{-1}$, $h_1,h_2 \in H$. Then $W : H \times H \to \gamma_2(G)$ is a 2-cocycle and satisfies

(W_1) $\quad W(\lambda_1^{r_1}\lambda_2^{r_2} \ldots \lambda_n^{r_n}, \lambda_1^{s_1}\lambda_2^{s_2} \ldots \lambda_n^{s_n}) = \prod_{1 \leq i \leq j \leq n} W(\lambda_i^{r_i}, \lambda_j^{s_j})$

(W_2) $\quad W(a,b)$, $a \in C(\lambda_1) \oplus \ldots \oplus C(\lambda_i)$, $b \in C(\lambda_{i+1}) \oplus \ldots \oplus C(\lambda_n)$ is bilinear for $i = 1,2,\ldots,n-1$.

Let $1 \to T \to M \to G \to 1$ be a fixed central extension of T by G . We shall write the groups T and M additively and the group G multiplicatively.

6.1 Lemma. *It is possible to choose a set of representatives $\{u(g)\}_{g \in G}$ of G in M such that*

$$f(g_1,g_2) = u(g_1)u(g_2)u(g_1,g_2)^{-1} , \ g_1,g_2 \in G ,$$

satisfies the following conditions:

(f_1) $\quad f(z_1,z_2) = 0$, $z_1,z_2 \in \gamma_2(G)$;

(f_2) $\quad f(z,w(h)) = 0$, $z \in \gamma_2(G)$, $h \in H$;

(f_3) $\quad f(w(\lambda_i^r),w(\lambda_i^s)) = 0$, $i = 1,2,\ldots,n$ *and* r,s *arbitrary;*

(f_4) $\quad f(w(b),w(a)) = 0$, $a \in C(\lambda_1) \oplus \ldots \oplus C(\lambda_i)$, $b \in C(\lambda_{i+1}) \oplus \ldots \oplus C(\lambda_n)$, $i = 1,2,\ldots,n-1$

Proof. Since G is of class 2 and

$$1 \to T \to M \to G \to 1$$

is a central extension, the nilpotency class of M is ≤ 3 . Therefore, the subgroup $\gamma_2(M)T$, which is the inverse image of $\gamma_2(G)$, is Abelian. Now T is divisible Abelian. Therefore, the exact sequence

$$1 \to T \to \gamma_2(M)T \to \gamma_2(G) \to 1$$

splits. Hence we can choose representatives $\{u(z)\}_{z\in\gamma_2(G)}$ of elements of $\gamma_2(G)$ in M such that (f_1) is satisfied.

Let m_i be an arbitrary representative of $w(\lambda_i)$, where i is any one of the integers $1,2,\ldots,n$. Then $-u(w(\lambda_i)^{p^{\alpha_i}}) + p^{\alpha_i}m_i$ is an element of T. Therefore, we can find an element $t_i \in T$ such that

(6.2) $\quad -u(w(\lambda_i)^{p^{\alpha_i}}) + p^{\alpha_i}m_i = p^{\alpha_i}t_i$.

Choose

(i) $\quad u(w(\lambda_i)) = m_i - t_i$, $i = 1,2,\ldots,n$;

(ii) $\quad u(w(\lambda_i^r)) = ru(w(\lambda_i))$, $0 \leq r < p^{\alpha_i}$;

(iii) $\quad u(w(\lambda_1^{r_1}\lambda_2^{r_2}\ldots\lambda_n^{r_n})) = u(w(\lambda_n^{r_n})) + u(w(\lambda_{n-1}^{r_{n-1}}))+\ldots+u(w(\lambda_1^{r_1}))$.

(iv) $\quad u(zw(h)) = u(z) + u(w(h))$, $z \in \gamma_2(G)$, $h \in H$.

Then (iv) assures (f_2) and (iii) assures (f_4) .
Let $0 \leq r, s < p^{\alpha_i}$. Then

$$u(w(\lambda_i^r)) + u(w(\lambda_i^s)) = u(w(\lambda_i^r)w(\lambda_i^s)) + f(w(\lambda_i^r),w(\lambda_i^s)) .$$

(i) and (ii) give

$$r(m_i-t_i) + s(m_i-t_i) = u(w(\lambda_i^{r+s})W(\lambda_i^r,\lambda_i^s)) + f(w(\lambda_i^r),w(\lambda_i^s)) .$$

Therefore, by (iv)

$$(r+s)(m_i-t_i) = u(W(\lambda_i^r,\lambda_i^s)) + u(w(\lambda_i^{r+s})) + f(w(\lambda_i^r),w(\lambda_i^s))$$

$$= \begin{cases} u(w(\lambda_i)^{p^{\alpha_i}}) + (r+s-p^{\alpha_i})(m_i-t_i) \\ +f(w(\lambda_i^r),w(\lambda_i^s)) \quad \text{if} \quad r+s \geq p^{\alpha_i} , \\ (r+s)(m_i-t_i) + f(w(\lambda_i^r),w(\lambda_i^s)) \quad \text{if} \quad 0 \leq r,s < p^{\alpha_i} . \end{cases}$$

Thus $f(w(\lambda_i^r),w(\lambda_i^s)) = 0$ (use (6.2) and observe that t_i is in the centre of M). Hence $f(w(\lambda_i^r),w(\lambda_i^s)) = 0$ for arbitrary integers r,s .

Assume that a choice of representatives $\{u(g)\}_{g\in G}$ has been made such that the corresponding 2-cocycle $f : G \times G \to T$ satisfies the conditions (f_i) , $i = 1,2,3,4$. We extend f by linearity to $Z(G) \times Z(G)$. The conditions (f_i), $i = 1,2,3,4$, and the 2-cocycle condition

$$f(x-1,(y-1)(z-1)) = f((x-1)(y-1),z-1) , \quad x,y,z \in G$$

satisfied by f imply that

(6.3) $f(w(h_1)z_1, w(h_2)z_2) = f(w(h_1), w(h_2)) + f(w(h_1), z_2)$,
$(h_1, h_2 \in H \; ; \; z_1, z_2 \in \gamma_2(G))$

and that the function

(6.4) $k(h,z) = f(w(h),z) \; (h \in H, z \in \gamma_2(G))$

is bilinear on $H \times \gamma_2(G)$.

Let

(6.5) $\bar{f}(h_1,h_2) = f(w(h_1),w(h_2)), h_1,h_2 \in H$.

If $g_i = w(h_i)z_i$, $i = 1,2,3,4$, then (6.3) and (6.4) imply that

$$f((g_1-1)(g_2-1)(g_3-1),g_4) = \bar{f}((h_1-1))h_2-1)(h_3-1),h_4) ,$$

where $\bar{f}$ has been extended by linearity to $Z(H) \times Z(H)$.

The equation

$$f((w(h_1)-1)(w(h_2)-1),w(h_3)) = f(w(h_1),(w(h_2)-1)(w(h_3)-1))$$

implies that

(6.6) $\bar{f}((h_1-1)(h_2-1),h_3) = \bar{f}(h_1,(h_2-1)(h_3-1)) + k(h_1,W(h_2,h_3))$,
$h_1,h_2,h_3 \in H$.

If $a_1,a_2 \in C(\lambda_1)$ and $b_1,b_2 \in \sum_{i=2}^{n} \oplus C(\lambda_i)$, then making use of (6.1) and (6.6), we have

$$\bar{f}(a_1b_1,a_2b_2) = k(b_1,W(a_1,a_2)) + k(b_2,W(a_1,a_2)) + k(b_1,W(a_1,b_2)) + \bar{f}(a_1,b_2) + \bar{f}(b_1,b_2) .$$

Thus if $h_i = a_ib_i$, $a_i \in C(\lambda_1)$, $b_i \in \sum_{j=2}^{n} \oplus C(\lambda_j)$, $i = 1,2,3,4$, then writing $\bar{x} = x - 1$ for $x \in H$, we have

$$\begin{aligned}\bar{f}(\bar{h}_1\bar{h}_2\bar{h}_3,h_4) &= k(b_1,W(a_1\bar{a}_2\bar{a}_3,a_4)) + k(b_2,W(a_2\bar{a}_1\bar{a}_3,a_4)) \\ &+ k(b_3,W(a_3\bar{a}_1\bar{a}_2,a_4)) + k(b_4,W(a_4\bar{a}_1\bar{a}_2,a_3)) \\ &+ \bar{f}(\bar{b}_1\bar{b}_2\bar{b}_3,b_4) .\end{aligned}$$

By iterating the above process, we obtain

$$\begin{aligned}\bar{f}(\bar{h}_1\bar{h}_2\bar{h}_3,h_4) = \sum_{1 \leq j<i \leq n} \{ & k(\lambda_i^{r_{i1}},W(\lambda_j^{r_{j1}}\lambda_j^{\bar{r}_{j2}}\lambda_j^{\bar{r}_{j3}},\lambda_j^{r_{j4}})) \\ &+ k(\lambda_i^{r_{i2}},W(\lambda_j^{r_{j2}}\lambda_j^{\bar{r}_{j3}}\lambda_j^{\bar{r}_{j1}},\lambda_j^{r_{j4}})) \\ &+ k(\lambda_i^{r_{i3}},W(\lambda_j^{r_{j3}}\lambda_j^{\bar{r}_{j1}}\lambda_j^{\bar{r}_{j2}},\lambda_j^{r_{j4}}))\end{aligned}$$

$$+\ k(\lambda_i^{r_{i4}}, W(\lambda_j^{r_{j4}}\lambda_j^{\bar{r}_{j1}}\lambda_j^{\bar{r}_{j2}}, \lambda_j^{r_{j3}}))$$

where $h_i = \lambda_1^{r_{1i}}\lambda_2^{r_{2i}} \dots \lambda_n^{r_{ni}} \in H$, $i = 1,2,3,4$.

Define $\chi : G \to T$ by

$$\chi(w(\lambda_1^{r_1}\lambda_2^{r_2}\dots\lambda_n^{r_n})z) = \sum_{i=1}^{n} k(\lambda_i^{r_i}, z) +$$

$$\sum_{1\leq j<i\leq n} \{k(\lambda_i^{r_i}, W(\lambda_j, \lambda_j+\lambda_j^2+\dots+\lambda_j^{r_j-1})) - k(\lambda_j^{r_j}, W(\lambda_i, \lambda_i+\lambda_i^2+\dots+\lambda_i^{r_i-1}))$$

$$+\ r_i r_j \bar{f}(\lambda_i, \lambda_j) .$$

Then it can be shown that

$2f((g_1-1)(g_2-1)(g_3-1), g_4) = \chi((g_1-1)(g_2-1)(g_3-1)(g_4-1))$, $g_i \in G$, $i = 1,2,3,4$.

Thus the 2-cocycle $\beta(g_1,g_2) = 2f(g_1,g_2) - \chi((g_1-1)(g_2-1))$ is a polynomial 2-cocycle of degree ≤ 2 . Therefore, the cohomology class of the 2-cocycle β lies in $P_2H^2(G,T)$. In view of the well-known correspondence between the elements of $H^2(G,T)$ and the central extensions of T by G , we have proved

6.7 Theorem. *If* G *is a finite* p-*group of class* 2, *then* $2H^2(G,T) \subseteq P_2H^2(G,T)$.

In particular, for odd prime power groups we have

6.8 Theorem [63]. *If* G *is a finite* p-*group of class* 2, $p \neq 2$, *then* $P_2H^2(G,T) = H^2(G,T)$.

We can now compute the fourth integral dimension subgroup for p-groups, $p \neq 2$.

6.9 Theorem [67]. *If* Π *is a* p-*group*, $p \neq 2$, *then*

$$\gamma_4(\Pi) = \Pi \cap (1+\Delta_Z^4(\Pi) + \Delta_Z(\Pi)\Delta_Z(\zeta_1(\Pi)))$$

where $\zeta_1(\Pi)$ *is the centre of* Π .

Proof. It is clearly enough to prove the result for finite p-groups of class 3, $p \neq 2$. Let Π be such a group. Then $\gamma_4(\Pi) = 1$. If possible, let $1 \neq x \in \Pi \cap (1+\Delta_Z^4(\Pi) + \Delta_Z(\Pi)\Delta_Z(\zeta_1(\Pi)))$.
Then, by Theorem 5.9, $x \in \gamma_3(\Pi) \subseteq \zeta_1(\Pi)$. We can find a homomorphism $\alpha : \zeta_1(\Pi) \to T$ such that $\alpha(x) \neq 0$. Since $P_2H^2(\Pi/\zeta_1(\Pi),T) = H^2(\Pi/\zeta_1(\Pi),T)$ (Theorem 6.8), the homomorphism α can be extended to

a map $\varphi : \to T$ whose linear extension to $Z(\Pi)$ vanishes on $\Delta_Z^4(\Pi) + \Delta_Z(\Pi)\Delta_Z(\zeta_1(\Pi))$ (Theorem 5.1). But then $\alpha(x) = \varphi(x) = \varphi(x-1) = 0$, a contradiction. Hence $\Pi \cap (1+\Delta_Z^4(\Pi) + \Delta_Z(\Pi)\Delta_Z(\zeta_1(\Pi))) = 1$ and the Theorem is proved.

6.10 <u>Corollary</u> [63]. <u>If</u> Π <u>is a</u> p-<u>group</u>, $p \neq 2$, <u>then</u>

$$D_4(\Pi) = \gamma_4(\Pi) .$$

For a long time it was conjectured that all groups have the dimension property, i.e. for every group G, $D_n(G) = \gamma_n(G)$ for all $n \geq 1$. This conjecture was refuted in 1969 by Rips [78] who constructed a 2-group G of order 2^{38} and nilpotency class 3 ($\gamma_4(G) = 1$) such that $D_4(G) \neq 1$. Tahara ([95], [96]) has now constructed for all integers $k \geq 2$ and $\ell \geq 0$ a 2-group G of order $2^{8k+\ell+22}$ and class 3 such that $D_4(G) \neq 1$. In particular, when $k = 2$ and $\ell = 0$, Tahara's group is the same as the group constructed by Rips.

It is still an open problem whether every nilpotent group must have its integral dimension series terminating with identity after a finite number of steps. Even the weaker question - " Is $\bigcap_n D_n(G)$ always equal to $\bigcap_n \gamma_n(G)$?" - is also not yet settled (see [76] for these problems).

Though $D_4(\Pi)$ may fail to equal $\gamma_4(\Pi)$, yet $D_4(\Pi)/\gamma_4(\Pi)$ has exponent at most 2. This has been shown by Losey [45], Sjögren [90] and Tahara [94] and can also be deduced easily from the treatment given here.

6.11 <u>Theorem</u>. <u>For every group</u> Π, $D_4(\Pi)/\gamma_4(\Pi)$ <u>has exponent at most 2</u>.

<u>Proof</u>. It suffices to verify this result for finite 2-groups of class 3 (see 4.4 for the necessary reductions involved). Let Π be a finite 2-group of class 3. If possible, let $x \in D_4(\Pi)$ and $x^2 \neq 1$. Then $x \in \gamma_3(\Pi)$ (Theorem 5.10) and there exists a homomorphism $\alpha : \gamma_3(\Pi) \to T$ such that $2\alpha(x) = \alpha(x^2) \neq 0$. Since $P_2H^2(\Pi/\gamma_3(\Pi),T) \supseteq 2H^2(\Pi/\gamma_3(\Pi),T)$ (Theorem 6.7), the homomorphism $2\alpha : \gamma_3(\Pi) \to T$ can be extended to a polynomial map $\varphi : \Pi \to T$ of degree ≤ 3 (Theorem 5.1). We thus have $2\alpha(x) = \varphi(x) = \varphi(x-1) = 0$, a contradiction. Hence $D_4(\Pi) = 1$ and the Theorem is proved.

Tahara has recently computed the fourth and fifth integral dimension subgroups of finite groups ([95], [97]). We give his result for the fourth integral dimension subgroup without proof.

Let $G = \gamma_1(G) \supseteq \gamma_2(G) \supseteq \gamma_3(G) \supseteq \gamma_4(G) = 1$ be the lower central series of a finite group G. Let x_{1i} $(1 \leq i \leq s)$ be the elements of G such that $\bar{x}_{1i} = x_{1i}\gamma_2(G)$ $(1 \leq i \leq s)$ are a basis of the finite Abelian group $G/\gamma_2(G)$, $d(i)$ is the order of $\bar{x}_{1i}$ in $\gamma_1(G)/\gamma_2(G)$ and $d(i) \mid d(i+1)$. We put for $1 \leq i \leq s$

$$x_{1i}^{d(i)} = x_{21}^{c_{i1}} x_{22}^{c_{i2}} \cdots x_{2t}^{c_{it}} x_{3i}, \text{ where } x_{2j}\gamma_3(G), \ 1 \leq j \leq t,$$

form a basis of $\gamma_2(G)/\gamma_3(G)$ and $x_{3i} \in \gamma_3(G)$. Let $d'(k)$ = order of $x_{2k}\gamma_3(G)$.

6.12 Theorem [95]. $D_4(G)$ is equal to the subgroup of G generated by the elements

$$\prod_{1 \leq i < j \leq s} (x_{1i}^{d(i)}, x_{1j})^{u_{ij} \frac{d(j)}{d(i)}}$$

for all integers u_{ij}, $1 \leq i < j \leq s$ satisfying the following conditions:

$$(6.13) \qquad u_{ij}\binom{d(j)}{2} \equiv 0 \pmod{d(i)}$$

and for $1 \leq i \leq s$, $1 \leq k \leq t$,

$$(6.14) \qquad \sum_{1 \leq h < i} u_{hi} \frac{d(i)}{d(h)} c_{hk} - \sum_{i < j \leq s} u_{ij} c_{jk} \equiv 0 \pmod{(d(i), d'(k))}.$$

In an earlier paper [94], Tahara has shown that $D_4(G) = \gamma_4(G)$ if G is a finite 2-group of class 3 and either $G/\gamma_2(G)$ is a direct sum of at most three cyclic groups or $\gamma_2(G)$ is cyclic (see [63] for the case when $G/\gamma_2(G)$ is a direct sum of at most two cyclic groups).

Finally, we mention that Sjögren [90] has recently published the following

6.15 Theorem [90]. Let b_m = least common multiple $\{1, 2, \ldots, m\}$, $a_1^n = 1$, $a_{k+1}^n = \prod_{i=1}^{n} a_i^{n-k+i} b_{n-k}$. If $c_n = \prod_{k=1}^{n-1} a_k^k$, then $D_n^{c_n}(G) \subseteq \gamma_n(G)$.

The first few c_i's are as follows:

$c_1 = 1$, $c_2 = 1$, $c_3 = 1$, $c_4 = 2$, $c_5 = 48$.

It is easy to see that if p is a prime, then p does not divide c_n for $1 \leq n \leq p+1$. We thus have the following Corollary which improves a result of Moran [56] and includes 6.10.

6.16 Corollary [90]. Any p-group G satisfies $D_n(G) = \gamma_n(G)$ for $n \leq p+1$.

Sjögren's approach involves the comparison of the series of quotients γ_n/γ_{n+1} with that of Δ^n/Δ^{n+1} by means of a spectral sequence. The detailed argument uses Fox derivatives [13] and methods of Chen, Fox and Lyndon [10] and involves calculations of logrithms in a formal setting (the numbers a_i^n enter through the denominators). Moran's original proof is also based on calculation with formal logrithms and we give this in the next Section.

7. MODULAR DIMENSION SUBGROUPS

The Campbell-Hausdorff formula is defined in the context of free Lie rings over the field $\mathbb{Q}$ of rational numbers by means of the formal power series for log and exp as

$$x o y = \log((\exp x)(\exp y)).$$

Dynkin [16] expanded this as follows:

$$(7.1) \qquad x o y = \sum_k \frac{(-1)^{k-1}}{k} \sum \frac{\psi(x^{a_{11}} y^{a_{21}} \ldots x^{a_{1k}} y^{a_{2k}})}{\Pi(a_{ij}!)}$$

where the sum is over all $a_{ij} \geqslant 0$, $1 \leqslant i \leqslant 2$, $1 \leqslant j \leqslant k$, such that $a_{1j} + a_{2j} > 0$ for all j and where

$$\psi(y_1 y_2 \ldots y_n) = \frac{1}{n}[\ldots[y_1, y_2], \ldots, y_n] .$$

Let L be a Lie ring of characteristic p^h, p prime, $h \geqslant 1$. Suppose that L is nilpotent of class $c < p$. Then the formula (7.1) defines a multiplication on L because p appears in the denominator of a coefficient of a Lie bracket only if the Lie bracket has weight $\geqslant p$, when, by assumption, it is zero. With this multiplication, which is associative, L becomes a monoid. The zero element of L acts as the multiplicative identity. Let M denote this monoid (L, o).

7.2 <u>Theorem</u>. <u>The map</u> $\mu : M \to L$ <u>given by</u> $\mu(x) = x$, $x \in M$, <u>is a polynomial map of degree</u> $\leqslant c$. (See proof of Theorem 3.15 [79].)

<u>Proof</u>. Extend μ to $Z(M)$ by linearity. Then we have to prove that μ vanishes on $\Delta_Z^{c+1}(M)$. It is clearly enough to prove that μ vanishes on $(g_1-1)(g_2-1)\ldots(g_{c+1}-1)$, $g_i \in M$, $i = 1,2,\ldots,c+1$. We have in $Z(M)$,

$$(7.3) \qquad (g_1-1)(g_2-1)\ldots(g_{c+1}-1) = \Sigma(-1)^{c+1-s} g_{\alpha_1} o g_{\alpha_2} o \ldots o g_{\alpha_s}$$

where the summation is taken over all distinct subsets $\{\alpha_1, \alpha_2, \ldots, \alpha_s\}$ of $\{1,2,\ldots,c+1\}$ with $1 < \alpha_1 < \alpha_2 < \ldots < \alpha_s \leqslant c+1$.
The image γ of $(g_1-1)(g_2-1)\ldots(g_{c+1}-1)$ under μ has the same expression as the right hand side of (7.3) but $g_{\alpha_1} o g_{\alpha_2} o \ldots o g_{\alpha_s}$ is to be expanded using (7.1). Now

$$x_1 \circ x_2 \circ \dots \circ x_n = \sum_k \frac{(-1)^k}{k} \sum \frac{\psi(x_1^{a_{11}} x_2^{a_{21}} \dots x_n^{a_{n1}} \dots x_1^{a_{1k}} x_2^{a_{2k}} \dots x_n^{a_{nk}})}{\Pi(a_{ij}!)}$$

where the sum is over all $a_{ij} \geq 0$, $1 \leq i \leq n$, $1 \leq j \leq n$, such that $\sum_i a_{ij} > 0$ for all j. Let S be a proper subset of $\{1,2,\dots,c+1\}$. Choose integers $a_{ij} \geq 0$, $1 \leq i \leq c+1$, $1 \leq j \leq k$ such that $a_{sj} = 0$ for all j if s is not in S. Then the term

$$w(a_{ij}) = \frac{(-1)^{k-1} \psi(g_1^{a_{11}} \dots g_{c+1}^{a_{c+1,k}})}{k\Pi(a_{ij}!)}$$

does not involve the elements g_s for s not in S. The term $w(a_{ij})$ appears in the expansion of $g_{\alpha_1} \circ g_{\alpha_2} \circ \dots \circ g_{\alpha_s}$ if and only if S is a subset of $\{\alpha_1, \alpha_2, \dots, \alpha_s\}$ and when it appears, it appears with coefficient $(-1)^{c+1-s}$. Thus, in the expansion for γ, $w(a_{ij})$ appears with coefficient $\Sigma(-1)^{c+1-s}$, the sum over all subsets of s elements which contain S. Hence the coefficient of $w(a_{ij})$ is

$$\sum_{s=t}^{c+1} (-1)^{c+1-s} \binom{c+1-t}{s-t} = (1-1)^{c+1-t} = 0 ,$$

since $c+1 > t$. Finally, note that in the expansion of γ, the term involving all the g_i's is zero. For, by assumption, L is nilpotent of class c so that any Lie commutator of weight $c+1$ is zero. Hence it follows that $\gamma = 0$ i.e. μ vanishes on $\Delta_Z^{c+1}(M)$.

7.4 Corollary. *If* $x - 1_M \in \Delta_Z^{c+1}(M)$, *then* $x = 1_M$, *where* 1_M *is the identity of* M.

7.5 Theorem [56]. *Let* G *be a* p-*group*. *Then*

$$D_{n,Z/p^hZ}(G) = G^{p^h} \gamma_n(G)$$

for $n = 1,2,\dots,p$, $h \geq 1$.

Proof. Since $p^h | \binom{p^h}{i}$ for $i = 1,2,\dots,p-1$, the identity

$$x^{p^h} - 1 = \sum_{i=1}^{p^h} \binom{p^h}{i} (x-1)^i$$

shows that $G^{p^h} \subseteq D_{p,Z/p^hZ}(G)$. Also $\gamma_n(G) \subseteq D_{n,Z/p^hZ}(G)$ for all $n \geq 1$. Hence $G^{p^h}\gamma_n(G) \subseteq D_{n,Z/p^hZ}(G)$ for $n = 1,2,\dots,p$. Suppose equality holds for $n = 1,2,\dots,c$ $(c < p)$ and $G^{p^h}\gamma_{c+1}(G) \neq D_{c+1,Z/p^hZ}(G)$. Then we can pick a finite subquotient H of G such that $H^{p^h} = 1$, $\gamma_{c+1}(H) = 1$, but $D_{c+1,Z/p^hZ}(H) \neq 1$ (note that $D_{c+1,Z/p^hZ}(H) = D_{c+1,Z}(H)$, a consequence of $H^{p^h} = 1$) . By a theorem of Lazard [38], H is isomorphic to (L,o) for some nilpotent Lie ring L of class < c and characteristic p^h. Hence, by Corollary 7.4, $D_{c+1,Z/p^hZ}(H) = 1$, a contradiction. Hence $D_{n,Z/p^hZ}(G) = G^{p^h}\gamma_n(G)$ for $n = 1,2,\ldots,p$.

7.6 <u>Corollary</u> [56]. <u>Any</u> p-<u>group</u> G <u>satisfies</u>

$$D_n(G) = \gamma_n(G)$$

<u>for</u> $n = 1,2,\ldots,p$.

<u>Proof</u>. Let G be a p-group. Suppose $D_n(G) = \gamma_n(G)$ for $n = 1,2,\ldots,c$ $(c < p)$ and $D_{c+1}(G) \neq \gamma_{c+1}(G)$. Then we can find a finite subquotient H of G such that $D_{c+1}(H) \neq \gamma_{c+1}(H)$. Let p^h be the order of H . If $x \in D_{c+1}(H)$, then $x \in D_{c+1,Z/p^hZ}(H) = H^{p^h}\gamma_{c+1}(H) = \gamma_{c+1}(H)$, showing that $D_{c+1}(H) \subseteq \gamma_{c+1}(H)$, a contradiction. Hence $D_n(G) = \gamma_n(G)$ for $n = 1,2,\ldots,p$.

7.7 <u>Remarks</u>. (i) Modular dimension subgroups $D_{n,Z/p^hZ}(F)$ have been computed by Lazard [38] for all n and $h \geqslant 1$ for free groups F . These subgroups are analogous to the subgroups $G_{n,p}$ of Chapter IV, Section 1. Their properties have been established group theoretically, by an analogous use of Dark's theorem, by Sandling [80]. The formula for the free case is not correct for the case of p-groups except for 7.5 as has been shown by Moran [56] and Sandling [80].
(ii) It follows from Theorem 7.5 that $D_{2,Z/p^hZ}(G) = G^{p^h}\gamma_2(G)$ (as is easy to see directly (Chapter II, Example 1.5)) and $D_{3,Z/p^hZ}(G) = G^{p^h}\gamma_3(G)$ for odd primes p . The subgroup $D_{3,Z/2^hZ}(G)$ has been shown by Sandling [80] to be equal to $\langle x^{2^h} \mid x^{2^{h-1}} \in G^{2^h}\gamma_2(G)\rangle\gamma_3(G)$ (see also [72] for the computation of the third modular dimension subgroup).

CHAPTER VI

THE RESIDUAL NILPOTENCE OF THE AUGMENTATION IDEAL

In this Chapter we study the nilpotence and the residual nilpotence of the augmentation ideal of a group ring.

1. THE NILPOTENCE OF THE AUGMENTATION IDEAL

1.1 Definition. An ideal I of a ring R is said to be *nilpotent* if $I^n = 0$ for some $n \geq 1$.

It was shown by Jennings [35] and by Lombardo-Radice [41] that $\Delta_K(G)$ is nilpotent if G is a finite p-group and K is a field of characteristic $p \neq 0$. The converse of this was established by Losey [43]: if $\Delta_K(G)$ is nilpotent, then G is a finite p-group. The complete answer to the nilpotence of the augmentation ideal of an arbitrary group ring is given by the following result.

Let G be a non-identity group and R a ring with identity (not necessarily commutative).

1.2 Theorem [12]. *The augmentation ideal $\Delta_R(G)$ is nilpotent if and only if G is a finite p-group and p is nilpotent in R.*

Proof. Suppose $\Delta_R^n(G) = 0$ and $\Delta_R^{n-1}(G) \neq 0$, $n \geq 2$. Let $0 \neq z = \sum_{\tau_x \in R, x \in G} \tau_x x \in \Delta_R^{n-1}(G)$. Replacing z by $y^{-1}z$, where $y \in G$ is such that $\tau_y \neq 0$, if necessary, we can assume that τ_1, the coefficient of identity in z, is non-zero. Since $\Delta_R^n(G) = 0$, we have

$$(g-1)z = 0 \quad \text{for all} \quad g \in G .$$

Now in gz, $g \in G$, the coefficient of g is τ_1. Hence it follows that

$$\tau_g = \tau_1 \quad \text{for all} \quad g \in G .$$

Thus G must be finite. To show that G must be a prime power group, it is clearly enough to prove that G cannot have elements of coprime orders. If possible, let x and y be elements of orders r and s respectively, where $1 \neq r$, $1 \neq s$ and g.c.d. $(r,s) = 1$. The equation

$$r(x-1) = -\{\binom{r}{2}(x-1)^2 + \ldots + \binom{r}{r-1}(x-1)^{r-1} + (x-1)^r\}$$

shows that

$$r(x-1) \in \Delta_R^2(\langle x\rangle) ,$$

where $\langle x \rangle$ is the subgroup generated by x. Hence

$$r^{n-1}(x-1) \in \Delta_R^n(\langle x\rangle) \subseteq \Delta_R^n(G) = 0 .$$

Similarly

$$s^{n-1}(x-1) \in \Delta_R^n(\langle y\rangle) \subseteq \Delta_R^n(G) = 0 .$$

Since g.c.d. $(r^{n-1}, s^{n-1}) = 1$, there exist integers u and v such that

$$1 = r^{n-1}u + s^{n-1}v.$$

Consequently

$$(x-1)(y-1) = (r^{n-1}u + s^{n-1}v)(x-1)(y-1) = 0 .$$

But (x, y being of coprime order) $(x-1)(y-1) \neq 0$. This contradiction establishes our assertion that G must be a prime power group.

Suppose G is of order p^m, p prime, $m \geq 1$. We need to show that p must be nilpotent in R. Let $1 \neq g \in G$. As above

$$p^{m(n-1)}(g-1) \in \Delta_R^n(G) = 0 .$$

Hence

$$p^{m(n-1)} = 0 \text{ in } R ,$$

proving that p is nilpotent in R.

Conversely, suppose G is a finite p-group and p is nilpotent in R. We first show that

(1.3) $$\Delta_Z^m(G) \subseteq p\Delta_Z(G) \quad \text{for some } m \geq 2.$$

This is clear if G is Abelian. For example, if order of G is p^n, then the identity

$$x^{p^n} - 1 = \sum_{i=1}^{p^n} \binom{p^n}{i}(x-1)^i$$

shows that

$$(g-1)^{p^n} \in p\Delta_Z(G) \quad \text{for all } g \in G$$

and, therefore,

$$\Delta_Z^{(p^n-1)^2+1}(G) \subseteq p\Delta_Z(G) .$$

Suppose (1.3) holds for finite p-groups of class $< c$ and G has class c $(c > 1)$. If $\gamma_c(G)$ is the last non-identity term of the lower central series of G, then $G/\gamma_c(G)$ is of class $c - 1$ and, by induction hypothesis,

$$\Delta_Z^m(G/\gamma_c(G)) \subseteq p\Delta_Z(G/\gamma_c(G)) \quad \text{for some} \quad m \geq 2 .$$

This is equivalent to saying that

$$\Delta_Z^m(G) \subseteq p\Delta_Z(G) + \Delta_Z(\gamma_c(G))\cdot Z(G) ,$$

since $\Delta_Z(\gamma_c(G))\cdot Z(G)$ is the kernel of the natural map $Z(G) \to Z(G/\gamma_c(G))$. From the Abelian case, there exists m' such that

$$\Delta_Z^{m'}(\gamma_c(G)) \subseteq p\Delta_Z(\gamma_c(G)) \subset p\Delta_Z(G) .$$

Hence

$$\Delta_Z^{mm'}(G) \subseteq p\Delta_Z(G)$$

and (1.3) is established. Iteration of (1.3) shows that

$$\Delta_Z^{s(m-1)+1}(G) \subseteq p^s\Delta_Z(G) \quad \text{for every} \quad s \geq 1.$$

Thus if $p^s = 0$ in R, then

$$\Delta_R^{s(m-1)+1}(G) = 0$$

i.e. $\Delta_R(G)$ is nilpotent.

We next consider the vanishing of a Lie power $\Delta_R^{(n)}(G)$ of $\Delta_R(G)$ (R commutative) (see Chapter I, (1.2) for definition of $\Delta_R^{(n)}(G)$). If G is Abelian, then trivially $\Delta_R^{(2)}(G) = 0$ and conversely. The following Theorem characterizes the non-Abelian case.

1.4 Theorem [59]. Let G be a non-Abelian group, R a commutative ring with identity. Then $\Delta_R^{(n)}(G) = 0$ for some $n > 2$ if and only if G is nilpotent, $\gamma_2(G)$ is a finite p-group and p is nilpotent in R. [$\gamma_i(G)$ denotes the i-th term in the lower central series of G].

Proof. Suppose $\Delta_R^{(n)}(G) = 0$ for some $n > 2$. Then $D_{(n),R}(G)$ $(= G \cap (1+\Delta_R^{(n)}(G))) = 1$ and so G is nilpotent (note that $\{D_{(i),R}(G)\}_{i\geq 1}$ is an N-series (Chapter IV, Proposition 2.1) and, therefore, $\gamma_i(G) \subseteq D_{(i),R}(G)$ for all $i \geq 1$). Also $(\Delta_R^{(2)}(G))^{n-1} \subseteq \Delta_R^{(n)}(G) = 0$ (Chapter I, Proposition 1.7). As $\Delta_R(\gamma_2(G)) \subseteq \Delta_R^{(2)}(G)$, we have

$$\Delta_R^{n-1}(\gamma_2(G)) = 0.$$

Therefore, by Theorem 1.2, $\gamma_2(G)$ is a finite p-group and p is nilpotent in R.

Conversely, suppose G is a nilpotent group and $\gamma_2(G)$ is a finite p-group of order p^r, say $(r \geq 1)$. Then $\gamma_n(G) = 1$ and $\Delta_R^n(\gamma_2(G)) = 0$ for sufficiently large n. Consequently, $\Delta_R^{(n^2)}(G) = 0$ (Chapter I, Theorem 1.8).

2. THE RESIDUAL NILPOTENCE OF THE AUGMENTATION IDEAL

2.1 Definition. An ideal I of a ring R is said to be *residually nilpotent* if $\bigcap_n I^n = 0$.

For convenience, we adopt the following

2.2 Notation $I^{\omega} = \bigcap_n I^n$

We are interested in the residual nilpotence of the augmentation ideals $\Delta_R(G)$ of group rings $R(G)$. We begin with the work of Jennings [36] on finitely generated torsion-free nilpotent groups. Jennings' result states that $\Delta_{\mathbb{Q}}^{\omega}(G) = 0$ if G is finitely generated torsion-free nilpotent and $\mathbb{Q}$ is the field of rational numbers. Formanek [18] has given an elegant short proof of this result and has noted that Jennings' Theorem holds for arbitrary coefficient rings. It may be pointed out that Formanek's observation is contained in Theorem B2 of Hartley [27].

Let R be a ring (not necessarily commutative) with identity. Let

$$1 \to H \to G \to Z \to 1$$

be an exact sequence of (multiplicative) groups (Z = infinite cyclic group). If t is a generator of Z and $x \in G$ is an element which maps onto the generator t, then $R(H)$ can be regarded as a left $R(G)$-module by defining the action of H on $R(H)$ to be left multiplication and the action of x on H to be

$$x \cdot h = xhx^{-1} .$$

This action, of course, depends on the choice of x.

2.3 Swan's Lemma [93]. *Let*

$$1 \to H \to G \to Z \to 1$$

be an exact sequence with G *nilpotent. Let* $R(H)$ *be regarded as a*

<u>left</u> R(G)-<u>module as above</u>. <u>Then for each integer</u> m

$$\Delta_R^{m^c}(G)\cdot R(H) \subseteq \Delta_R^m(H) \quad ,$$

<u>where</u> c <u>is the nilpotency class of</u> G.

<u>Proof</u>. If $c = 1$, i.e. G is Abelian, then the sequence splits and $G = H \oplus Z$. The assertion in this case holds trivially.

Suppose G is nilpotent of class $c > 1$ and the result holds for nilpotent groups of class $c - 1$. Let $A = \gamma_c(G)$, the c-th term in the lower central series of G. Then

$$A \subseteq \gamma_2(G) \subseteq H \ .$$

By induction,

$$\Delta_R^{m^{c-1}}(G/A) . R(H/A) \subseteq \Delta_R^m(H/A) .$$

or

$$\Delta_R^{m^{c-1}}(G)\cdot R(H) \subseteq \Delta_R^m(H) + R(H)\cdot\Delta_R(A) .$$

Since A is in the centre of G, R(H) is a left R(G) and right R(A) bimodule. Hence

$$\begin{aligned}\Delta_R^{2m^{c-1}}(G) . R(H) &\subseteq \Delta_R^m(H) + \Delta_R^{m^{c-1}}(G)\cdot(R(H)\cdot\Delta_R(A))\\ &\subseteq \Delta_R^m(H) + (\Delta_R^m(H) + R(H)\cdot\Delta_R(A))\Delta_R(A)\\ &\subseteq \Delta_R^m(H) + R(H)\cdot\Delta_R^2(A) \ .\end{aligned}$$

This argument eventually gives

$$\Delta_R^{m^c}(G)\cdot R(H) \subseteq \Delta_R^m(H)$$

which completes the induction.

2.4 <u>Definition</u>. A group G is called an <u>R-group</u> if it has the following property:

$$\text{"}x,y \in G,\ x^i = y^i \text{ for some } i \geqslant 1 \rightarrow x = y\text{"}.$$

2.5 <u>Lemma</u>. <u>A torsion-free nilpotent group is an R-group</u>.

<u>Proof</u>. We proceed by induction on the nilpotency class. The result is obvious for Abelian groups. Let G be a torsion-free nilpotent group of class $n > 1$ and assume that the Lemma holds for torsion-free

nilpotent groups of class < n. Let $x,y \in G$ and $x^i = y^i$ for some $i > 0$. If Z_1 is the centre of G, then G/Z_1 is torsion-free nilpotent of class < n. Now in G/Z_1, $(xZ_1)^i = (yZ_1)^i$. Therefore, by induction hypothesis, $xZ_1 = yZ_1$, i.e. $x = yz$ for some $z \in Z_1$. Since $x^i = y^i$, we have $z^i = 1$ (z being in the centre of G). As G is torsion-free, z must be 1. Hence $x = y$. This completes the induction and the Lemma is proved.

2.6 Lemma. Let

$$1 \to H \to G \to Z \to 1$$

be an exact sequence with $H \neq 1$ and G an R-group. Suppose $\alpha \in R(G)$ and $\alpha \neq 0$. Then there exists an $x \in G$ mapping onto a generator $t \in Z$ such that

$$\alpha \cdot R(H) \neq 0$$

where the module action of $R(G)$ on $R(H)$ is defined relative to x.

Proof. Let $\alpha = \sum_{\tau_x \in R, x \in G} \tau_x x$. We may assume without loss of generality that τ_1, the coefficient of identity in α, is not zero. Let

$$\text{Support of } \alpha = \{1, g_1, g_2, \ldots, g_r\}$$

(i.e. the set of elements $x \in G$ for which $\tau_x \neq 0$). If $x,y \in G$ map onto $t \in Z$ and $x \neq y$, then $\langle x \rangle \cap \langle y \rangle = 1$ ($\langle z \rangle$ denotes the subgroup generated by z). For, if $x^i = y^j$, then $t^i = t^j$ and, therefore, $i = j$. Since G is an R-group, we have $x = y$. Thus each of $g_1, g_2, \ldots, g_r$ lies in at most one such subgroup $\langle x \rangle$. Since H is infinite, there are infinitely many $x \in G$ mapping onto $t \in Z$, so we can choose one with $g_1, g_2, \ldots, g_r \notin \langle x \rangle$. Then

$$g_1 = h_1 x^{i_1}, \ldots, g_r = h_r x^{i_r}$$

where $h_1, \ldots, h_r \in H$, $h_1, \ldots, h_r \neq 1$ and i_j's are integers. Consider $1 \in R(H)$.

$$1 \cdot 1 = 1, \quad g_s \cdot 1 = h_s x^{i_s} \cdot 1 = h_s \neq 1$$

for $s = 1, 2, \ldots, r$. Therefore, $\alpha \cdot 1 \neq 0$ as required.

2.7 Theorem. Let G be a finitely generated torsion-free nilpotent group and R any ring with identity. Then $\Delta_R(G)$ is residually nilpotent.

Proof. It is well-known that a finitely generated torsion-free nilpotent group G possesses a series

$$G = H_1 \supseteq H_2 \supseteq \cdots \supseteq H_n \supseteq H_{n+1} = 1$$

such that H_i is a normal subgroup of H_{i-1} and H_{i-1}/H_i is infinite cyclic for $i = 1,2,\ldots,n+1$. [Such a series can be obtained, for example, by refining the upper central series $\{Z_i\}$ of G.] The length of such a series is an invariant of G and is called the torsion-free rank (or Hirsch number) of G. We proceed by induction on $n \geq 1$. If $n = 1$, then G is infinite cyclic and $R(G) \cong R[t,t^{-1}]$, the ring of polynimals in t and t^{-1}. Under this isomorphism $\Delta_R^r(G)$ is mapped onto the ideal generated by $(t-1)^r$ and the result is clear.

Let $n > 1$ and

$$1 \to H \to G \to Z \to 1$$

be an exact sequence with rank of $H = n - 1$ (for example, take $H = H_2$) so that $\Delta_R^\omega(H) = 0$ by induction hypothesis. Let $\alpha \in \Delta_R^\omega(G)$. If $\alpha \neq 0$, then, by Lemma 2.6, we can find an element $x \in G$ which maps onto t, a generator of Z, so that under the $R(G)$ action on $R(H)$ defined relative to x, $\alpha\cdot R(H) \neq 0$. However, by Swan's Lemma,

$$\alpha\cdot R(H) \subseteq \Delta_R^\omega(H) = 0 .$$

This is a contradiction and Theorem 2.7 is proved.

The proof of the next result depends on the theory developed in Chapter III. For any unexplained notation or terminology the reader may refer to Chapter III.

2.8 Theorem [27]. Let G be a group having a finite N-series

$$G = G_1 \supseteq G_2 \supseteq \cdots \supseteq G_c \supseteq G_{c+1} = 1$$

each factor G_i/G_{i+1}, $i \geq 1$, of which is the direct product of a free Abelian group and a group of exponent dividing p^K, where p is a fixed prime and K is a fixed integer. Let R be a ring with identity. Then $\Delta_R(G)$ is residually nilpotent provided

either the factors G_i/G_{i+1}, $i \geq 1$, are all torsion-free

or $\bigcap_n p^n R = 0$

Proof. Let $(x_\alpha)_{\alpha<\lambda}$ be a canonical basis for G (Chapter III, (2.1)). Then for every choice of maps $g : N \to N$ (N = the set of natural numbers), $\nu : R\setminus\{0\} \to N \cup \{0\}$ satisfying (Chapter III, (2.11) and

(2.12)) and integer M, we can construct R-submodules E_n, $n \geq 0$, as in (Chapter III, (2.14)). Since either the quotients G_i/G_{i+1}, $i \geq 1$, are all torsion-free or $\bigcap_n p^n R = 0$, Theorem 2.15 of Chapter III applies. Thus there exists a choice of g and ν such that the resulting R-submodules E_n satisfy

$$E_r \cdot E_s \subseteq E_{r+s} \quad \text{for all} \quad r,s \geq 0 .$$

Each of these E_n's then is an ideal of $R(G)$. We observe that E_n's are independent of the choice of the integer M used in their construction. More specifically, if M and M' are integers $\geq n$ and $E_{n,M}$, $E_{n,M'}$ are constructed relative to M and M' respectively, then

$$E_{n,M} = E_{n,M'} .$$

For

$$(1-x_\beta)^M \in E_{n,M'} \quad \text{for all} \quad \beta < \lambda$$

and, since $E_{n,M'}$ is an ideal, any product of elements of $R(G)$ involving $(1-x_\beta)^M$ lies in $E_{n,M'}$. Hence

$$E_{n,M} \subseteq E_{n,M'}$$

and the equality follows by symmetry. In the following we shall always mean by E_n this "asymptotic" E_n. As $\Delta_R^n(G) \subseteq E_n$ for all $n \geq 1$, it is enough to prove the following

2.9 Lemma. $\bigcap_n E_n = 0$

Proof. If x is an arbitrary element of G, we can write x in its canonical form with respect to the canonical basis $(x_\alpha)_{\alpha<\lambda}$ of G as

$$x = \prod_{\alpha<\lambda} x_\alpha^{f_\alpha(x)}$$

where $f_\alpha(x) \in Z$. Observe that if $u_1, u_2, \ldots, u_n$ are finitely many elements of G, then there exists $y \in G$ such that

$$f_\alpha(u_i y) \geq 0$$

for all $\alpha < \lambda$ and $i = 1,2,\ldots,n$. This assertion can be proved by induction on the nilpotency class of G.

Let $u = \sum_{\tau_x \in R, x \in G} \tau_x x$ be an element of $\bigcap_n E_n$ and suppose that not all τ_x are zero. Since each E_n is an ideal of $R(G)$, we have $uy \in E_n$ for all $y \in G$. Hence, in view of the preceding observation, we can assume that $f_\alpha(x) \geq 0$ for all $\alpha < \lambda$ and all x such that

$\tau_x \neq 0$. Since $u \in E_n$, we can write (for a suitably chosen M)

$$u = \Sigma\lambda(\underset{\sim}{r})u(\underset{\sim}{r}) \quad , \; \underset{\sim}{r} \in \underset{\sim}{T}$$

where $\underset{\sim}{T}$ is a non-empty finite set of vectors of $\underset{\sim}{S}$ (see Chapter III for definitions and notations), $\lambda(\underset{\sim}{r})$'s are non-zero elements of R and $\nu(\lambda(\underset{\sim}{r})) + \nu(u(\underset{\sim}{r})) \geq n$ for all $\underset{\sim}{r} \in \underset{\sim}{T}$. By Lemma 2.9 of Chapter III, there exists $\underset{\sim}{s} = (s_\alpha) \in \underset{\sim}{T}$ such that when u is expressed as an R-linear combination of elements of G , the element $\prod_{\alpha<\lambda} x_\alpha^{s_\alpha}$ occurs with coefficient $\pm\lambda(\underset{\sim}{s})$. Thus $s_\alpha \geq 0$ for all $\alpha < \lambda$ and

$$\nu(\lambda(\underset{\sim}{s})) + \nu(u(\underset{\sim}{s})) \leq \operatorname{Max}_x \nu(\tau_x) + \sum_{\alpha<\lambda} \nu(\alpha) \max_x f_\alpha(x) = \ell \quad ,$$

a number depending only on u . If $n > \ell$, we have a contradiction. Hence $u = 0$ and the Lemma is proved. This completes the proof of Theorem 2.8.

2.10 <u>Notations</u>. (i) $\bar{\mathfrak{N}}_p$ = <u>the class of nilpotent</u> p-<u>groups of bounded exponent</u> [a group G is said to be of <u>bounded exponent</u> if there exists an integer $m \geq 1$ such that $x^m = 1$ for all $x \in G$].

(ii) If $\mathfrak{C}$ is a class of groups, then $\mathfrak{R}\mathfrak{C}$ denotes the class of groups G which satisfy:

$1 \neq x \in G \Rightarrow$ there exists a normal subgroup N_x of G such that $x \notin N_x$ and $G/N_x \in \mathfrak{C}$.

It may be noted that the class $\bar{\mathfrak{N}}_p$ consists precisely of those groups G whose dimension series $\{D_{n,Z/pZ}(G)\}_{n\geq1}$ terminates with identity after a finite number of steps. We shall need the observation that the class $\bar{\mathfrak{N}}_p$ is closed under finite direct products.

2.11 <u>Theorem</u>. <u>If</u> $G \in \mathfrak{R}\bar{\mathfrak{N}}_p$ <u>and</u> R <u>is a ring with identity satisfying</u> $\bigcap_n p^nR = 0$, <u>then</u> $\Delta_R(G)$ <u>is residually nilpotent</u>.

<u>Proof</u>. Let

$$G = \gamma_1(G) \supseteq \gamma_2(G) \supseteq \cdots \supseteq \gamma_c(G) \supseteq \gamma_{c+1}(G) = 1$$

be the lower central series of a group $G \in \bar{\mathfrak{N}}_p$. Then the series $\{\gamma_i(G)\}$ satisfies the requirements of Theorem 2.8 (use [37], Theorem 6). Hence $\Delta_R(G)$, where R satisfies $\bigcap_n p^nR = 0$, is residually nilpotent. Thus the Theorem holds for the groups in the class $\bar{\mathfrak{N}}_p$. Next let $G \in \mathfrak{R}\bar{\mathfrak{N}}_p$ and

$$\sigma = \sum_{i=1}^{n} r_i x_i \; , \qquad (r_i \in R \; , \; x_i \in G \; , \; x_i\text{'s all distinct})$$

belong to $\Delta_R^{\omega}(G)$. Since $G \in \mathfrak{R}\bar{\mathfrak{N}}_p$, there exist normal subgroups N_{ij} of G such that $G/N_{ij} \in \bar{\mathfrak{N}}_p$ and $x_i^{-1}x_j \notin N_{ij}$, $1 \leq i < j \leq n$. Consider the homomorphism

$$\theta : G \to \prod_{1\leq i<j\leq n} (G/N_{ij})$$

given by

$$\theta(g) = (gN_{ij}).$$

Clearly

$$\text{Ker } \theta = \bigcap_{1\leq i<j\leq n} N_{ij} = N, \text{ say.}$$

Then G/N is isomorphic to a subgroup of the direct product

$$P = \prod_{1\leq i<j\leq n} (G/N_{ij}) .$$

Since the class $\bar{\mathfrak{N}}_p$ is closed under finite direct products, $P \in \bar{\mathfrak{N}}_p$. Thus $\Delta_R^{\omega}(P) = 0$. Hence $\Delta_R^{\omega}(G/N) = 0$. Let $\varrho : R(G) \to R(G/N)$ be the R-linear extension of the natural projection of G onto G/N. Since ϱ is a ring homomorphism and $\varrho(\Delta_R(G)) \subseteq \Delta_R(G/N)$, $\varrho(\Delta_R^{\omega}(G)) \subseteq \Delta_R^{\omega}(G/N) = 0$. Hence $\sum_{i=1}^{n} r_i(x_iN) = 0$. Since, by the choice of N_{ij}, $x_iN \neq x_jN$ for $i \neq j$, it follows that each r_i is zero. Consequently $\varrho = 0$. Hence $\Delta_R^{\omega}(G) = 0$.

2.12 Notation. Let p be a prime. We denote by $\mathfrak{K}_p$ <u>the class of those nilpotent groups whose derived groups are p-groups of bounded exponent</u>.

It may be observed that the class $\mathfrak{K}_p$ consists of those groups G whose Lie dimension series $\{D_{(n),Z/pZ}(G)\}_{n\geq 1}$ terminates with identity in a finite number of steps. Clearly the class $\mathfrak{K}_p$ is closed under finite direct products.

2.13 Theorem [59]. If $G \in \mathfrak{R}\mathfrak{K}_p$ and R is a commutative ring with identity satisfying $\bigcap_n p^nR = 0$, then $\bigcap_n \Delta_R^{(n)}(G) = 0$.

Proof. As the class $\mathfrak{K}_p$ is closed under finite direct products, it clearly suffices to prove that

$$\text{' } G \in \mathfrak{K}_p \Rightarrow \bigcap_n \Delta_R^{(n)}(G) = 0 \text{ '} .$$

For a nilpotent group G, Theorem 1.8 of Chapter I shows that

$$\bigcap_n \Delta_R^{(n)}(G) \subset \bigcap_m (\Delta_R^m(\gamma_2(G))\cdot R(G)) .$$

Let $R(G)$ be regarded as a left $R(\gamma_2(G))$-module with $\gamma_2(G)$ acting on $R(G)$ by left multiplication. Let $(t_i)_{i\in I}$ be a set of coset representatives of $\gamma_2(G)$ in G. Then every element $z \in R(G)$ can be written <u>uniquely</u> as

$$(*) \qquad z = \sum_{i\in I} \alpha(i) t_i$$

$\alpha(i) \in R(\gamma_2(G))$, $\alpha(i)$ being zero except for a finite number of i's. If $z \in \Delta_R^m(\gamma_2(G))\cdot R(G)$, then we can express z as

$$z = \sum_{i\in I} \beta(i) t_i$$

with $\beta(i) \in \Delta_R^m(\gamma_2(G))$. Hence, from uniqueness of the expression (*), $\alpha(i) \in \Delta_R^m(\gamma_2(G))$ for all $i \in I$. It follows that

$$\bigcap_m (\Delta_R^m(\gamma_2(G))\cdot R(G)) \subset (\bigcap \Delta_R^m(\gamma_2(G)))\cdot R(G) .$$

Since $G \in \mathfrak{R}_p$, $\gamma_2(G) \in \bar{\mathfrak{N}}_p$. Hence, by Theorem 2.11 $\bigcap_m \Delta_R^m(\gamma_2(G)) = 0$ Consequently

$$\bigcap_n \Delta_R^{(n)}(G) = 0 .$$

2.14 <u>Torsion-free nilpotent groups</u>

According to Theorem 2.7 $\Delta_R(G)$ is residually nilpotent for every finitely generated torsion-free nilpotent group G and any ring R with identity. This result is <u>not</u> true for arbitrary torsion-free nilpotent groups. For, let G be a divisible group and $0 \neq r \in R$, $0 < n \in \mathbb{Z}$ be such that $nr = 0$. Then, for $g \in G$,

$$\begin{aligned} r(g-1) &= r(h^n-1) \qquad \text{(since } G \text{ is divisible there exists } h \in G \text{ such that } g = h^n) \\ &= r\{n(h-1) + \tbinom{n}{2}(h-1)^2 + \dots + (h-1)^n\} \\ &\equiv 0 \bmod r\Delta_R^2(G) \quad \text{(since } nr = 0). \end{aligned}$$

Hence $r\Delta_R(G) = r\Delta_R^2(G)$. Consequently $r\Delta_R(G) \subseteq \Delta_R^\omega(G)$. Thus a necessary condition on R for the residual nilpotence of $\Delta_R(G)$, G arbitrary torsion-free nilpotent, is that the additive group $(R,+)$ should be torsion-free. We prove that this condition is in fact, sufficient.

2.15 Theorem If G is a residually 'torsion-free nilpotent group' and R is a ring with identity such that its additive group $(R,+)$ is torsion-free, then $\Delta_R(G)$ is residually nilpotent.

In order to prove Theorem 2.15 we need information about the way the group ring $R(K)$ of a subgroup K of a finitely generated torsion free nilpotent group H intersects with the powers $\Delta_R^r(H)$, $r \geq 1$, of the augmentation ideal $\Delta_R(H)$ of $R(H)$.

Let H be a finitely generated torsion-free nilpotent group of class c, say, and R a ring with identity. Then H has a finite N-series

(2.16)
$$H = H_1 \supseteq H_2 \supseteq \dots \supseteq H_c \supseteq H_{c+1} = 1$$

such that the quotients H_i/H_{i+1}, $i \geq 1$, are all torsion-free. We can take, for example, the series $\{D_{i,\mathbb{Q}}(G)\}_{i\geq 1}$ of dimension subgroups of H over $\mathbb{Q}$, the field of rational numbers. Choose a canonical basis $\{x_1,x_2,\dots,x_N\}$ for H adapted to the series (2.16) and let R-submodules $\bar{E}_n$, $n \geq 0$, be constructed as in Chapter III. We assume that each $\bar{E}_n$ is asymptotic in the sense explained in the proof of Theorem 2.8. Recall that, by definition, $\bar{E}_n$ is spanned, as an R-submodule of $R(G)$, by the elements $u(\underset{\sim}{r})$ of weight $\geq n$. Here $\underset{\sim}{r} = (r_1,r_2,\dots,r_N)$ is a vector of integers and

(2.17)
$$u(\underset{\sim}{r}) = v_1 v_2 \dots v_N ,$$

where, writing $u_i = 1-x_i$,

$$v_i = \begin{cases} u_i^{r_i}, & r_i \geq 0 \\ u_i^M x_i^{r_i}, & r_i < 0 \end{cases}$$

M being an integer $\geq n$. The weight of $u(\underset{\sim}{r})$ is taken to be $\sum_{i=1}^{N} r_i \mu_i$, where μ_i is defined by $x_i \in H_{\mu_i} \setminus H_{\mu_i+1}$, if all r_i 's are ≥ 0 and M otherwise. Since Theorem 2.15 of Chapter III is applicable, each

$\bar{E}_n$, $n \geq 0$, is an ideal of $R(G)$.

2.18 Lemma $\bar{E}_{nc} \subseteq \Delta_R^n(H)$ *for all* $n \geq 1$.

Proof. Let $u(\underset{\sim}{r})$ be an element of the form (2.17) having weight $\geq nc$ $(M \geq nc)$. If some $r_i < 0$, then the product $u(\underset{\sim}{r})$ contains $u_i^M \in \Delta_R^n(H)$. If all $r_i \geq 0$, then the weight of $u(\underset{\sim}{r})$ is $\sum_{i=1}^N r_i\mu_i$. As $\mu_i \leq c$ for all i, $\sum_{i=1}^N r_i\mu_i \geq nc$ only if $\sum_{i=1}^N r_i \geq n$. Hence $u(\underset{\sim}{r}) \in \Delta_R^n(H)$.

The following Lemma is the main step in the proof of Theorem 2.15.

2.19 Lemma. *If* H *is a finitely generated torsion-free nilpotent group of class* c, K *is a subgroup of* H *and* R *is a ring with identity such that* $(R,+)$ *is torsion-free, then*

$$\Delta_R^{rc}(H) \cap R(K) \subseteq \Delta_R^r(K)$$

for all $r \geq 1$.

Proof. The quotients H_i/H_{i+1}, $i \geq 1$, of the series (2.16) are all free Abelian of finite rank. Let $K_i = H_i \cap K$, $i \geq 1$. Then $\{K_i\}_{i\geq 1}$ is a finite N-series of K and

$$K_i/K_{i+1} \cong K_iH_{i+1}/H_{i+1}, \quad i \geq 1.$$

By the fundamental theorem ([29], Theorem 7.1) on free Abelian groups, there exists a Z-basis $H_{i+1}w_1,\ldots,H_{i+1}w_\ell$ of H_i/H_{i+1} and strictly positive integers $s_1,\ldots,s_k$, $k \leq \ell$, such that the cosets $H_{i+1}w_j^{s_j}$ form a Z-basis of $K_i\,H_{i+1}/H_{i+1}$. Write $w_j^{s_j} = y_j'y_j$, $y_j' \in H_{i+1}$, $y_j \in K_i$. Then the cosets $K_{i+1}y_j$, $1 \leq j \leq k$, form a Z-basis of K_i/K_{i+1}. In this manner we can choose a canonical basis $x_1,x_2,\ldots,x_N$ of H adapted to the series (2.16) and a canonical basis $\{y_i\}_{i\in S}$ indexed by a subset S of $\{1,2,\ldots,N\}$ and adapted to the N-series $\{K_i\}$ so that

$$y_i \equiv x_i^{s_i} \pmod{H_{\mu_i+1}}, \qquad i \in S$$

for some integer $s_i > 0$. Here μ_i is determined by $x_i \in H_{\mu_i} \setminus H_{\mu_i+1}$. Note that $y_i \in K_{\mu_i} \setminus K_{\mu_i+1}$. Suppose $z \in \Delta_R^{rc}(H) \cap R(K)$ for some $r \geq 1$.

Then $z \in \bar{E}_{rc}$ (recall that $\Delta_R^n(H) \subseteq \bar{E}_n$ for all $n \geq 0$). Since $z \in R(K)$, we can write

$$z = \Sigma\theta(\underset{\sim}{t})u'(\underset{\sim}{t}) \tag{2.20}$$

where $\underset{\sim}{t}$ is a vector $(t_i)_{i\in S}$ of integers, $\theta(\underset{\sim}{t}) \in R$ and we may suppose $\theta(\underset{\sim}{t})$ are all non-zero. Also $u'(\underset{\sim}{t}) = \prod_{i\in S} v_i'$, where v_i' is either $(1-y_i)^{t_i}$ or $(1-y_i)^{M}y_i^{t_i}$ according as $t_i \geq 0$ or < 0 (M is an integer $\geq rc$). If we can show that $\Sigma t_i\mu_i \geq rc$ for all $\underset{\sim}{t}$ in (2.20) with $t_i \geq 0$ for all i, then $z \in \bar{E}_{rc}(K)$ and so, by Lemma 2.18, $z \in \Delta_R^r(K)$. Here $\bar{E}_n(K)$ denotes the R-submodule of $R(K)$ spanned by $u(\underset{\sim}{r})$, $\underset{\sim}{r} = (r_i)_{i\in S}$, of weight $\geq n$, where $u(\underset{\sim}{r})$ are defined in the usual manner for the basis $(y_i)_{i\in S}$ of K.

Now

$$y_i = y_i'x_i^{s_i} \quad (y_i' \in H_{\mu_i+1}).$$

Thus

$$1 - y_i = (1-x_i^{s_i}) + (1-y_i')x_i^{s_i} .$$

Since $1 - y_i' \in \Delta_R(H_{\mu_i+1}) \subseteq \bar{E}_{\mu_i+1}$ and $\bar{E}_{\mu_i+1}$ is an ideal of $R(H)$,

$$1 - y_i \equiv 1 - x_i^{s_i} \pmod{\bar{E}_{\mu_i+1}} .$$

Now

$$x_i^{s_i} \equiv 1 - s_i(1-x_i) \pmod{\bar{E}_{\mu_i+1}} .$$

Hence

$$1 - y_i \equiv s_i(1-x_i) \pmod{\bar{E}_{\mu_i+1}} .$$

Hence, for $t > 0$,

$$(1-y_i)^t \equiv s_i^t(1-x_i)^t \pmod{\bar{E}_{\mu_i+1}} . \tag{2.21}$$

Furthermore, (2.21) gives

$$(1-y_i)^M y_i^t \in \bar{E}_M \tag{2.22}$$

for any integer t. Thus, for any vector $\underset{\sim}{t} = (t_i)_{i\in S}$ with $t_i \geq 0$ for all i, we have

$$u'(\underset{\sim}{t}) \equiv \varrho(\underset{\sim}{t})u(\underset{\sim}{t}) \pmod{\bar{E}_{\nu(\underset{\sim}{t})+1}} \tag{2.23}$$

where $\varrho(t) = \prod s_i^{t_i}$ is a non-zero integer and $\nu(\underset{\sim}{t})$ is the weight function defined by taking $\nu(\theta) = 0$ for $0 \neq \theta \in R$ and g = identity (Chapter III, (2.13)). We are assuming that $\bar{E}_n = \bar{E}_M$ for $n \geq M$. Let r_0 be the minimum weight of a term occurring in (2.20) and suppose that $r_0 < rc$. Then, since $z \in \bar{E}_{rc}$, (2.23) gives

$$\Sigma\theta(\underset{\sim}{t})\varrho(\underset{\sim}{t})u(\underset{\sim}{t}) \equiv 0 \pmod{\bar{E}_{r_0+1}},$$

the sum being taken over all $\underset{\sim}{t}$ occurring in (2.20) with $\nu(\underset{\sim}{t}) = r_0$. Since $\theta(\underset{\sim}{t})\varrho(\underset{\sim}{t}) \neq 0$ (recall that $\theta(\underset{\sim}{t}) \neq 0$, $\varrho(\underset{\sim}{t}) \neq 0$ and $(R,+)$ is torsion-free), this contradicts the linear independence of the elements $u(\underset{\sim}{r})$ (Chapter III, Lemma 2.10). It, therefore, follows that, for each $\underset{\sim}{t}$ occurring in (2.20), $\nu(\underset{\sim}{t}) \geq rc$. Hence $z \in \bar{E}_{rc}(K) \subseteq \Delta_R^r(K)$, as required.

2.24 Proof of Theorem 2.15

Since the class of torsion-free nilpotent groups is closed under finite direct products, it suffices to prove the Theorem for torsion-free nilpotent groups (see the argument in the proof of Theorem 2.11). So let G be a torsion-free nilpotent group. Let $z = \sum_{\tau_x \in R, x \in G} \tau_x x \in \Delta_R^\omega(G)$. If K denotes the subgroup generated by the elements x for which $\tau_x \neq 0$, then $z \in \Delta_R^\omega(K)$. For, let $r \geq 1$ be given. If c is the nilpotency class of G, then $z \in \Delta_R^{rc}(G)$ and, therefore, there is a finitely generated torsion-free nilpotent group H of nilpotency class $\leq c$ such that $z \in \Delta_R^{rc}(H)$. We can assume without loss of generality that $H \supseteq K$, because K is finitely generated. Thus, by Lemma 2.19, $z \in \Delta_R^{rc}(H) \cap R(K) \subseteq \Delta_R^r(K)$. Hence $z \in \Delta_R^\omega(K)$. But K being finitely generated torsion-free nilpotent, $\Delta_R^\omega(K) = 0$ by Theorem 2.7. Hence $\Delta_R^\omega(G) = 0$.

2.25 Notation. For every group G and ring R with identity, we denote by $D_{\omega,R}(G)$ the intersection of the dimension series $\{D_{i,R}(G)\}_{i\geq 1}$ of G over R.

The foregoing results on the residual nilpotence of the augmentation ideals are strong enough to give a complete characterization when the coefficients are in a field.

2.26 Theorem. Let G be a group, k a field. Then $\Delta_k(G)$ is residually nilpotent if and only if

either G is residually 'torsion-free nilpotent' and characteristic of k is zero,

or G is residually 'nilpotent p-group of bounded exponent' and characteristic of k is $p > 0$.

Proof. Case I: Characteristic of k is zero. If G is residually 'torsion-free nilpotent', then $\Delta_R^{\omega}(G) = 0$ by Theorem 2.15.

Conversely, let $\Delta_k^{\omega}(G) = 0$. Then $D_{\omega,k}(G) = 1$. Since $G/D_{i,k}(G)$ is torsion-free nilpotent for all $i \geq 1$ (Chapter IV, Theorem 1.5), it follows that G is residually 'torsion-free nilpotent'.

Case II: Characteristic of k is $p > 0$.

If G is residually 'nilpotent p-group of bounded exponent', then $\Delta_R^{\omega}(G) = 0$ by Theorem 2.11.

Conversely, let $\Delta_k^{\omega}(G) = 0$. Then $D_{\omega,k}(G) = 1$. Since $G/D_{i,k}(G)$ is a nilpotent p-group of bounded exponent (Chapter IV, Theorem 1.9) for all $i \geq 1$, it follows that G is residually 'nilpotent p-group of bounded exponent'.

2.27 Remark. From the above proof it is clear that over a field k,

$$\Delta_k^{\omega}(G) = 0 \Leftrightarrow D_{\omega,k}(G) = 1 .$$

The following theorem characterizes the property "$\bigcap_n \Delta_k^{(n)}(G) = 0$" over fields. The proof is parallel to that of Theorem 2.26 and is omitted.

2.28 Theorem. Let G be a group, k a field. Then the following statements are equivalent:

(i) $\bigcap_n \Delta_k^{(n)}(G) = 0$;

(ii) $\bigcap_n D_{(n),k}(G) = 1$;

(iii) either G is residually 'nilpotent with derived group torsion-free' and characteristic of k is zero,

or G is residually 'nilpotent with derived group p-group of bounded exponent' and characteristic of k is $p > 0$.

We now consider the residual nilpotence of $\Delta_Z(G)$. Theorems 2.11 and 2.26 show that $\Delta_Z(G)$ is residually nilpotent if G is either residually 'nilpotent p-group of bounded exponent' or residually 'torsion-free nilpotent'. The problem of characterizing the groups G with $\Delta_Z(G)$ residually nilpotent has recently been solved by Lichtman [39]. To state the result we need the following

2.29 Definition. Let $\mathfrak{C}$ be a class of groups. A group G is said to be discriminated by $\mathfrak{C}$ if for every finite subset $g_1, g_2, \ldots, g_n$ of distinct elements of G, there exists a group $H \in \mathfrak{C}$ and a homomorphism $\varphi : G \to H$ such that $\varphi(g_i) \neq \varphi(g_j)$ for $i \neq j$.

Note that if a class $\mathfrak{C}$ is closed under subgroups and finite direct sums, then to say that $G \in \mathfrak{R}\mathfrak{C}$ is equivalent to saying that G is discriminated by $\mathfrak{C}$. For an example of a group G which is discriminated by the class $\mathfrak{C} = \bigcup_{p \text{ prime}} \bar{\mathfrak{N}}_p$ but does not belong to $\mathfrak{R}\bar{\mathfrak{N}}_p$ for any prime p, see Lichtman [40].

2.30 <u>Theorem</u> [39]. <u>Let</u> G <u>be a group</u>. <u>Then</u> $\Delta_Z(G)$ <u>is residually nilpotent if and only if either</u> G <u>is residually 'torsion-free nilpotent'</u> <u>or</u> G <u>is discriminated by the class of nilpotent</u> p_i-<u>groups</u>, $i \in I$, <u>of bounded exponents, where</u> $\{p_i | i \in I\}$ <u>is some set of primes</u>.

<u>Proof</u>. If G is residually 'torsion-free nilpotent', then $\Delta_{\mathbb{Q}}(G)$ is residually nilpotent, where $\mathbb{Q}$ is the field of rational numbers (Theorem 2.26). Therefore, in particular, $\Delta_Z(G)$ is residually nilpotent. Now suppose G is discriminated by the class of nilpotent p_i-groups, $i \in I$, of finite exponents. If possible, let $0 \neq \sum_{i=1}^{n} m_i(g_i - 1) \in \Delta_Z^r(G)$ for all $r \geq 1$, where $g_1, g_2, \ldots, g_n$ are distinct non-identity elements of G and $m_i \in Z$. By hypothesis there exists a nilpotent p-group H of bounded exponent and a homomorphism $\varphi : G \to H$ such that $\varphi(g_i)$, $i = 1,2,\ldots,n$ are distinct non-identity elements of H. As $\sum_{i=1}^{n} m_i(\varphi(g_i)-1) \in \bigcap_r \Delta_Z^r(H)$ and $\Delta_Z(H)$ is residually nilpotent (Theorem 2.11), we get $\sum_{i=1}^{n} m_i(\varphi(g_i)-1) = 0$, a contradiction. Hence $\Delta_Z(G)$ must be residually nilpotent.

For the converse, two cases arise

<u>Case I</u>: G <u>is torsion-free</u>.

If $D_{\omega,\mathbb{Q}}(G) = 1$, then clearly G is residually 'torsion-free nilpotent'. Suppose $D_{\omega,\mathbb{Q}}(G) \neq 1$. Let $1 \neq g \in D_{\omega,\mathbb{Q}}(G)$. Then, for every integer r, there exists an integer s such that $s(g-1) \in \Delta_Z^r(G)$. Our assertion in this case is that G must be discriminated by the class of nilpotent p_i-groups, $i \in I$, of bounded exponents for some set $\{p_i | i \in I\}$ of primes. Suppose G does not have this property. Then there must exist a finite set $g_1, g_2, \ldots, g_n$ of distinct non-identity elements such that whenever N is a normal subgroup of G with G/N a nilpotent p-group of bounded exponent, at least one of the g_i's lies in N. It is easy to see that $G/D_{r,Z/p^mZ}(G)$ is a nilpotent p-group of bounded exponent for all $r,m \geq 1$. It, therefore, follows that

$$(g_1-1)(g_2-1)\dots(g_n-1) \in \Delta_Z^r(G) + p^m\Delta_Z(G)$$

for all integers $r,m \geq 1$ and all primes p. Hence

$$(g-1)(g_1-1)(g_2-1)\dots(g_n-1) \in \Delta_Z^r(G) \quad \text{for all} \quad r \geq 1 .$$

Since $\Delta_Z(G)$ is residually nilpotent, we get

$$(g-1)(g_1-1)(g_2-1)\dots(g_n-1) = 0$$

which is not possible for a torsion-free nilpotent group (see the proof of Theorem 1.2). This establishes our assertion.

Case II: G has a periodic element $\neq 1$.

Let $1 \neq g \in G$ be an element of order p, p prime. Let $h \in \bigcap_{n,m} D_{n,Z/p^mZ}(G)$. Then we observe that $(g-1)(h-1) \in \Delta_Z^n(G)$ for all $n \geq 1$. Hence $(g-1)(h-1) = 0$ which forces h to be 1. Consequently $\bigcap_{n,m} D_{n,Z/p^mZ}(G) = 1$. As $G/D_{n,Z/p^mZ}(G)$ is a nilpotent p-group of bounded exponent for all $n,m \geq 1$, we conclude that G is residually 'nilpotent p-group of bounded exponent' and so is discriminated by the class of nilpotent p-groups of bounded exponents. (Note that the class of nilpotent p-groups of bounded exponents is closed under subgroups and finite direct sums).

2.31 Comment. Let F be a non-cyclic free group and let R be a normal subgroup of F. It can be seen via the Magnus representation [49] of F/R', R' = the derived group of R, that if the augmentation ideal $\Delta_Z(F/R)$ is residually nilpotent, then the group F/R' is residually nilpotent. Passi [68] has shown that R/R', when regarded as an F/R-module via conjugation in F, is a faithful module. A consequence ([55], [68]) of this is that if F/R' is residually nilpotent, then $\Delta_Z(F/R)$ is residually nilpotent. In other words, if $1 \to R \to F \to G \to 1$ is a non-cyclic free presentation of a group G, then the following are equivalent:

(i) F/R' is residually nilpotent;

(ii) $\Delta_Z(F/R)$ is residually nilpotent.

Thus Theorem 2.30 provides necessary and sufficient conditions for the residual nilpotence of F/R'. This problem was first raised by Gruenberg [20] who settled the case when F/R is finite. In this case the condition is that F/R be a prime power group. The problem was further investigated by Mital [55] who settled the case when F/R has a periodic element (see also Bovdi [6]). For other related consequences of the residual nilpotence of $\Delta_Z(G)$, see Hurley [33] and Gupta-Passi [22].

CHAPTER VII

THE NILPOTENT RESIDUE OF AN AUGMENTATION IDEAL

Let G be a group, R a ring with identity. In the last Chapter we have studied the residual nilpotence of the augmentation ideal $\Delta_R(G)$. We now turn to the precise calculation of $\Delta_R^\omega(G) = \bigcap_n \Delta_R^n(G)$, which we call the nilpotent residue of $\Delta_R(G)$. This seems in general to be a difficult problem. We discuss in this Chapter some of the cases in which the nilpotent residues have been calculated.

1. ELEMENTS OF INFINITE p-HEIGHT

1.1 Definition. Let G be a group, p a prime. An element $g \in G$ is called an element of infinite p-height if, for every $i,j \geq 1$, there exists $x \in G$, $y \in \gamma_i(G)$ such that $x^{p^j} = gy$, where $\gamma_i(G)$ is the i-th term in the lower central series of G.

We denote by $G(p)$ the set of all the elements of infinite p-height in G. Note that if G is Abelian, then $G(p) = \bigcap_j G^{p^j}$.

The set $G(p)$ can be identified in terms of dimension subgroups of G over Z/pZ. We need the following

1.2 Lemma (cf. [50], Lemma 2). *Let G be a nilpotent group of class c and m an integer ≥ 1. Then every product of m^c-th powers of elements of G is an m-th power of an element of G.*

Proof. We proceed by induction on the class of G. For groups of class 1, the result is obvious. Suppose G is a group of class $c > 1$ and the Lemma holds for groups of class $c-1$. Let $x_1, x_2, \dots, x_t \in G$. Then, by hypothesis,

$$x_1^{m^c} x_2^{m^c} \dots x_t^{m^c} = (x_1^m)^{m^{c-1}} (x_2^m)^{m^{c-1}} \dots (x_t^m)^{m^{c-1}}$$

$$= y^m \bmod \gamma_c(\langle x_1^m, x_2^m, \dots, x_t^m\rangle) \quad \text{for some } y \in G.$$

By elementary commutator calculus every element of $\gamma_c(\langle x_1^m, x_2^m, \dots, x_t^m\rangle)$ can be written as z^m for some $z \in \gamma_c(G) \subseteq$ centre of G. Hence

$x_1^{m^c} x_2^{m^c} \dots x_t^{m^c} = y^m z^m = (yz)^m$ and the Lemma is proved.

1.3 Proposition. *Let G be a group, p a prime. Then*

$$G(p) = \bigcap_{k,\ell} G^{p^k}\gamma_\ell(G) = \bigcap_{m,n} D_{n,Z/p^mZ}(G) = \bigcap_n D_{n,Z/pZ}(G) .$$

Proof. Let the four sets occurring in the statement be named A_1, A_2, A_3 and A_4 respectively from left to right. Then, by Definition, $A_1 \subseteq A_2$. That $A_2 \subseteq A_1$ follows from Lemma 1.2. To show that $A_2 \subseteq A_3$, let $m,n \geq 1$ be integers. Choose an integer k large enough to satisfy

$$p^m \mid \binom{p^k}{i} \quad \text{for} \quad i = 1,2,\ldots,n-1 .$$

Then the identity

$$x^{p^k} - 1 = \sum_{i=1}^{p^k} \binom{p^k}{i} (x-1)^i$$

shows that $G^{p^k} \subseteq D_{n,Z/p^mZ}(G)$. Since $\gamma_n(G) \subseteq D_{n,Z/p^mZ}(G)$, we have $G^{p^k}\gamma_n(G) \subseteq D_{n,Z/p^mZ}(G)$. It, therefore, follows that $A_2 \subseteq A_3$. The inclusion $A_3 \subseteq A_4$ is trivial. Since G/A_2 is residually 'nilpotent p-group of bounded exponent', it follows that $A_4 \subseteq A_2$ (Chapter VI, Theorem 2.11) and the proof is complete.

1.4 Notations. (i) Let $\bar{\mathfrak{N}}$ denote the class of nilpotent groups of bounded exponents.

(ii) For a group G, let $K(G)$ denote the intersection of all normal subgroups X of G such that $G/X \in \bar{\mathfrak{N}}$.

(iii) For a group G and integers $m,n \geq 1$, let $D_{n,m}(G) = D_{n,Z/mZ}(G)$. Note that if G is a p-group, then $K(G) = G(p)$.

1.5 Proposition. For every group G, $K(G) = \bigcap_{n,m} D_{n,m}(G)$.

Proof. For the inclusion '$K(G) \subseteq \bigcap_{n,m} D_{n,m}(G)$', observe that $G/D_{n,m}(G)$ is a nilpotent group of bounded exponent for all $n,m \geq 1$. The reverse inclusion is a consequence of the following

1.6 Lemma. *If $G \in \bar{\mathfrak{N}}$, then $\bigcap_{n,m} D_{n,m}(G) = 1$.*

Proof. Let $G \in \bar{\mathfrak{N}}$. Then $G = \Pi S_p$ (direct product of the Sylow p-subgroups S_p of G). If $1 \neq g \in \bigcap_{n,m} D_{n,m}(G)$, then, by projecting to the Sylow p-subgroups S_p, we get $1 \neq g_p \in \bigcap_{n,m} D_{n,m}(S_p)$ for some prime p. However, since S_p is a nilpotent p-group of bounded exponent, $\bigcap_n D_{n,p}(S_p) = 1$. Thus we have a contradiction. Hence $\bigcap_{n,m} D_{n,m}(G) = 1$.

The property $G = K(G)$ is easily characterized. We have the following

1.7 Proposition. For any group G, $G = K(G)$ if and only if $G/\gamma_2(G)$ is divisible Abelian.

Proof. If $G/\gamma_2(G)$ is divisible Abelian, then for every $m \geq 1$, $\Delta_{Z/mZ}(G) = \Delta^2_{Z/mZ}(G)$. Therefore, $\Delta_{Z/mZ}(G) = \Delta^{\omega}_{Z/mZ}(G)$. Consequently, $G \subseteq \bigcap_{n,m} D_{n,m}(G) = K(G)$. Conversely, if $G = K(G)$, then $G \subseteq D_{2,Z/pZ}(G) = G^p\gamma_2(G)$ for all primes p. Hence $G/\gamma_2(G)$ is divisible Abelian.

2. $\Delta^{\omega}_Z(G)$ IN PERIODIC GROUPS

2.1 Definitions

(i) Let G be a group. An element $g \in G$ is called a generalized periodic element if for every $n \geq 1$, there exists $m \geq 1$ such that $g^m \in \gamma_n(G)$.

(ii) An element $g \in G$ is called a generalized p-element if for every $n \geq 1$ there exists $m \geq 1$ such that $g^{p^m} \in \gamma_n(G)$.

Note that if g is a generalized periodic element, then for every $n \geq 1$, there exists $m = m(n) \geq 1$ such that $m(g-1) \in \Delta^n_Z(G)$ and if $g \in G$ is a generalized p-element, then there exists $m = m(n) \geq 1$ such that $p^m(g-1) \in \Delta^n_Z(G)$.

2.2 Notations. Let G be a group. We write

$P(G) = \{g \in G | g$ is a generalized periodic element$\}$

$G_{\omega p} = \{g \in G | g$ is a generalized p-element$\}$.

If R is a ring, we write R^+ for the additive group of R. Thus $R^+(p) = \bigcap_n p^n R$. By R^+_p we mean the set of p-elements of R^+. We write $D_{\omega,R}(G) = \bigcap_n D_{n,R}(G)$. Note that $P(G) = D_{\omega,k}(G)$ for any field k of characteristic zero.

2.3 Theorem [60]. Let G be a group and R a ring with identity. Then

$$\Delta_R(K(G))\Delta_R(P(G))R(G) + \sum_p \{\Delta_R(G(p))\Delta_R(G_{\omega p}) + R^+(p)\Delta_R(G_{\omega p}) + R^+_p\Delta_R(G(p))\}R(G)$$
$$+ \Delta_R(G,D_{\omega,R}(G)) \subseteq \Delta^{\omega}_R(G) .$$

Proof. Let $x \in K(G)$ and $y \in P(G)$. Then for every $n \geq 1$ there exists $m \geq 1$ such that $m(y-1) \in \Delta^n_Z(G)$. Since $x-1 \in \Delta^n_Z(G) + m\Delta_Z(G)$ for all $m,n \geq 1$ (Proposition 1.5), we have $(x-1)(y-1) \in \Delta^n_Z(G)$ for all $n \geq 1$.

Hence $(x-1)(x-1) \in \Delta_Z^{\omega}(G)$ and, therefore $\Delta_R(K(G))\Delta_R(P(G)) \subseteq \Delta_R^{\omega}(G)$.

Let $x \in G(p)$, $y \in G_{\omega p}$. Then $x-1 \in \Delta_Z^n(G) + p^m\Delta_Z(G)$ for all $m,n \geq 1$ (Proposition 1.3). For given $n \geq 1$, there exists $m \geq 1$ such that $p^m(y-1) \in \Delta_Z^n(G)$. Hence $(x-1)(y-1) \in \Delta_Z^{\omega}(G)$ and consequently we have $\Delta_R(G(p))\Delta_R(G_{\omega p}) \subseteq \Delta_R^{\omega}(G)$.

Let $x \in G_{\omega p}$ and $r \in R^+(p)$. Then $r \in p^mR$ for all $m \geq 1$ and for every $n \geq 1$ there exists m such that $p^m(x-1) \in \Delta_Z^n(G)$. Hence $r(x-1) \in \Delta_R^n(G)$ for all $n \geq 1$. This proves that $R^+(p)\Delta_R(G_{\omega p}) \subseteq \Delta_R^{\omega}(G)$.

Let $r \in R_p^+$ and suppose that $p^mr = 0$. Let $x \in G(p)$. Then $x-1 \in \Delta_Z^n(G) + p^m\Delta_Z(G)$ for all $n \geq 1$. Hence $r(x-1) \in \Delta_R^n(G)$ for all $n \geq 1$. Consequently $R_p^+\Delta_R(G(p)) \subseteq \Delta_R^{\omega}(G)$.

Finally, the inclusion $\Delta_R(G,D_{\omega,R}(G)) \subseteq \Delta_R^{\omega}(G)$ follows trivially from the definition of $D_{\omega,R}(G)$.

2.4 Lemma. If G is the direct product of its subgroups X_i $(i \in I)$ and if $\Delta_Z(X_i)\Delta_Z(X_j) \subseteq \Delta_Z^{\omega}(G)$ whenever $i \neq j$ then for every ring R with identity

$$\Delta_R^{\omega}(G) = \bigoplus_i \Delta_R^{\omega}(X_i) \oplus \sum_{i\neq j} \Delta_R(X_i)\Delta_R(X_j)R(G) .$$

Proof. The identity $xy-1 = x-1 + y-1 + (x-1)(y-1)$ shows that

$$\Delta_R(G) = \bigoplus_i \Delta_R(X_i) \oplus \sum_{i\neq j} \Delta_R(X_i)\Delta_R(X_j)R(G) .$$

Since $\Delta_R(X_i)\Delta_R(X_j) \subseteq \Delta_R^{\omega}(G)$ for $i \neq j$, the lemma follows.

2.5 Theorem. If G is periodic and $G \supsetneq K(G) = 1$, then for every ring R with identity

$$\Delta_R^{\omega}(G) = \sum_i \Delta_R(S_{p_i})R^+(p_i) \oplus \sum_{i\neq j} \Delta_R(S_{p_i})\Delta_R(S_{p_j})R(G)$$

where S_{p_i} are the Sylow p_i-subgroups of G .

Proof. Since $K(G)$ contains the intersection of the lower central series of G, G is residually nilpotent. Since G is also periodic we have $G = \prod_i S_{p_i}$. Now $\Delta_Z(S_{p_i})\Delta_Z(S_{p_j}) \subseteq \Delta_Z^{\omega}(G)$, $i \neq j$. Hence, by Lemma 2.4,

$$\Delta_R^{\omega}(G) = \bigoplus_i \Delta_R^{\omega}(S_{p_i}) \oplus \sum_{i\neq j} \Delta_R(S_{p_i})\Delta_R(S_{p_j})R(G) .$$

Since $K(G) = 1$, S_{p_i} is a residually nilpotent p_i-group of bounded exponent for each i. Hence by Theorem 2.11 of Chapter VI, $\Delta_R^{\omega}(S_{p_i}) \subseteq R^+(p_i)\Delta_R(S_{p_i})$. However, by Theorem 2.3,

$R^+(p_i)\Delta_R(S_{p_i}) \subseteq \Delta_R^\omega(S_{p_i})$. Hence $\Delta_R^\omega(S_{p_i}) = R^+(p_i)\Delta_R(S_{p_i})$ and the Theorem is proved.

2.6 Corollary. *If* $G/\bigcap_n \gamma_n(G)$ *is periodic, then for every ring* R *with identity*

$$\Delta_R^\omega(G) \subseteq \sum_i \Delta_R(G_{\omega p_i})R^+(p_i) + \sum_{i\neq j} \Delta_R(G_{\omega p_i})\Delta_R(G_{\omega p_j})R(G) + \Delta_R(G,K(G)) .$$

Proof. Let $M = \bigcap_n \gamma_n(G)$. If G/M is periodic, then $G/M = \prod_p S_p(G/M)$ (direct product of Sylow p-subgroups of G/M). But $S_p(G/M) = G_{\omega p}/M$ and it maps onto $S_p(G/K(G))$ under the natural homomorphism of G/M onto G/K. Hence, by Theorem 2.5,

$$\Delta_R^\omega(G/K(G)) = \{\sum_i \Delta_R(G_{\omega p_i})R^+(p_i) + \sum_{i\neq j} \Delta_R(G_{\omega p_i})\Delta_R(G_{\omega p_j})R(G) +$$
$$\Delta_R(G,K(G))\}/\Delta_R(G,K(G))$$

which proves the Corollary.

2.7 Theorem [21]. *If* $G/\bigcap_n \gamma_n(G)$ *is periodic, then*

$$\Delta_Z^\omega(G) = \Delta_Z(K(G))\Delta_Z(G) + \sum_{i\neq j} \Delta_Z(G_{\omega p_i})\Delta_Z(G_{\omega p_j})Z(G) + \Delta_Z(G,D_{\omega,Z}(G)) .$$

Proof. If $G/\bigcap_n \gamma_n(G)$ is periodic, then every element of G is clearly a generalized periodic element. Therefore, $\Delta_Z(K(G))\Delta_Z(G) \subseteq \Delta_Z^\omega(G)$ (Theorem 2.3). Since $G_{\omega p_i} \subseteq G(p_j)$ for $i \neq j$, the second term is contained in $\Delta_Z^\omega(G)$ again by Theorem 2.3. The third term is of course trivially contained in $\Delta_Z^\omega(G)$. Now, by Corollary 2.6, $\Delta_Z^\omega(G) \subseteq \sum_{i\neq j} \Delta_Z(G_{\omega p_i})\Delta_Z(G_{\omega p_j})Z(G) + \Delta_Z(G,K(G))$. Since every element α of $\Delta_Z(G,K(G))$ can be written as $\alpha = x-1+\beta$, $x \in G$, $\beta \in \Delta_Z(K(G))\Delta_Z(G) \subseteq \Delta_Z^\omega(G)$, it follows that $\Delta_Z^\omega(G)$ is contained in the right hand side.

3. $\Delta_R^\omega(G)$ IN FINITELY GENERATED NILPOTENT GROUPS

If G is a finitely generated group and R is any ring with identity, then the nilpotent residue $\Delta_R^\omega(G)$ is described by the following result due to Parmenter-Passi-Sehgal [60].

3.1 Theorem. *Let* G *be a finitely generated nilpotent group and* R *a ring with identity. Then*

$\Delta_R^\omega(G) = \sum_{i\neq j} \Delta_R(S_{p_i}(G))\Delta_R(S_{p_j}(G))R(G) + \sum_i R^+(p_i)\Delta_R(S_{p_i}(G))R(G)$, <u>where</u> $S_{p_i}(G)$ <u>are the Sylow subgroups of</u> G .

We recall the following Lemma of Parmenter-Sehgal [61] which can be proved by induction on n .

3.2 <u>Lemma</u>. <u>Let</u> R <u>be a ring with identity and let</u> $G_1 \times G_2 \times \ldots \times G_n$ <u>be a normal subgroup of</u> H <u>with each</u> G_i <u>normal in</u> H . <u>Then</u>

$$\Delta_R(H,G_1 \times G_2 \times \ldots \times G_{n-1}) \cap \Delta_R(H,G_1 \times \ldots \times G_{n-2} \times G_n) \cap \ldots \cap \Delta_R(H,G_2 \times \ldots \times G_n) =$$

$$\sum_{i\neq j} \Delta_R(H,G_i)\Delta_R(H,G_j)$$

[The above result is given in [61] for R commutative. However, the commutativity of the ring does not play any role and the same proof works for arbitrary rings.]

<u>Proof of Theorem 3.1</u>. We proceed by induction on the order of the torsion subgroup $\tau(G)$ of G . If $\tau(G) = 1$, then, by Theorem 2.7 of Chapter VI, $\Delta_R^\omega(G) = 0$ and we have nothing to prove. Let $\tau(G) \neq 1$ and suppose the Theorem holds for finitely generated nilpotent groups with torsion subgroups having order smaller than that of $\tau(G)$. Then, for some prime p , $S_p(G) \neq 1$ and, by induction hypothesis, the Theorem holds for $G/S_p(G)$. Since $S_{p_i}(G/S_p(G)) = S_{p_i}(G)S_p(G)/S_p(G)$ for $p_i \neq p$ and $S_p(G/S_p(G)) = 1$, we have

$$\Delta_R^\omega(G) \subseteq \sum_{i\neq j} \Delta_R(S_{p_i}(G))\Delta_R(S_{p_j}(G))R(G) + \sum_i R^+(p_i)\Delta_R(S_{p_i}(G))R(G) + \Delta_R(G,S_p(G)) .$$

Let $\alpha \in \Delta_R^\omega(G)$. Then $\alpha = \beta + \gamma$ where

$$\beta \in \sum_{i\neq j} \Delta_R(S_{p_i}(G))\Delta_R(S_{p_j}(G))R(G) + \sum_i R^+(p_i)\Delta_R(S_{p_i}(G))R(G) \subseteq \Delta_R^\omega(G)$$

and $\gamma \in \Delta_R(G,S_p(G))$. Let $S_{p'} = \prod_{p_i\neq p} S_{p_i}(G)$ (= the set of all the periodic elements whose order is not divisible by p). Then $G/S_{p'}$ is a residually finite p-group [19]. Therefore, $\Delta_R^\omega(G) \subseteq \Delta_R(G,S_{p'}) + R^+(p)\Delta_R(G)$ (Chapter VI, Theorem 2.11). Since $\gamma \in \Delta_R^\omega(G)$, we have $\gamma \in \Delta_R(G,S_{p'}) + R^+(p)\Delta_R(G)$. But $\gamma \in \Delta_R(G,S_p(G))$. Therefore, going mod $S_{p'}$, we conclude that $\gamma \in \Delta_R(G,S_{p'}) + R^+(p)\Delta_R(G,S_p(G))$.

Suppose $\gamma = u+v$, $u \in \Delta_R(G,S_{p'})$, $v \in R^+(p)\Delta_R(G,S_p(G))$. Then $u \in \Delta_R(G,S_{p'}) \cap \Delta_R(G,S_p(G)) = \Delta_R(G,S_p(G))\Delta_R(G,S_{p'})$ (Lemma 3.2). Hence

$$\gamma \in \Delta_R(G,S_{p'})\Delta_R(G,S_p(G)) + R^+(p)\Delta_R(G,S_p(G))$$

$$\subseteq \sum_{i\neq j} \Delta_R(S_{p_i}(G))\Delta_R(S_{p_i}(G))R(G) + \sum_i R^+(p_i)\Delta_R(S_{p_i}(G))R(G).$$

This proves that $\Delta_R^\omega(G)$ is contained in the right hand side of the asserted equality. The reverse inclusion can be verified directly or deduced from Theorem 2.3 by observing that in a nilpotent group $G_{\omega p} = S_p(G)$ for every prime p and that $S_{p_i}(G) \subseteq G(p_j)$ for $i \neq j$.

4. $\Delta_R^\omega(G)$ OVER FIELDS

The nilpotent residue $\Delta_R^\omega(G)$ of an arbitrary group G can be completely described when the coefficients are in a field.

4.1 <u>Theorem</u> [60]. <u>Let</u> G <u>be a group</u>, k <u>a field</u>. <u>Then</u>

(i) $\Delta_k^\omega(G) = \Delta_k(G,G(p))$ <u>if characteristic of</u> k is $p > 0$;

(ii) $\Delta_k^\omega(G) = \Delta_k(G,P(G))$ <u>if characteristic of</u> k <u>is zero</u>.

(See 1.1 and 2.2 for notations.)

<u>Proof</u>. (i) By Proposition 1.3, $G(p) \subset D_{\omega,k}(G)$. Therefore, $\Delta_k(G,G(p)) \subseteq \Delta^\omega(G)$. It follows from Proposition 1.3 that $G/G(p)$ is residually'nilpotent p-group of bounded exponent'. Hence $\Delta_R^\omega(G/G(p)) = 0$ (Chapter VI, Theorem 2.26). Therefore, $\Delta_R^\omega(G) \subseteq \Delta_R(G,G(p))$. This proves (i).

(ii) Clearly $P(G) = \bigcap_n \sqrt{\gamma_n(G)} = D_{\omega,k}(G)$ and $G/P(G)$ is residually 'torsion-free nilpotent'. Therefore, $\Delta_k(G,P(G)) \subseteq \Delta_k^\omega(G)$ and $\Delta_k^\omega(G) \subseteq \Delta_k(G,P(G))$ (Chapter VI, Theorem 2.26).

5. THE INTERSECTION THEOREM

Whenever an element $\alpha \in \Delta_R(G)$ satisfies the equation $\alpha = \alpha i$ for some $i \in \Delta_R(G)$, then evidently $\alpha \in \Delta_R^\omega(G)$. We now examine some cases where $\Delta_R^\omega(G)$ consists entirely of elements satisfying such an equation for a fixed $i \in \Delta_R(G)$, i.e. $\Delta_R(G)$ satisfies the "<u>inter-</u>

section theorem". We first consider a simple case.

5.1 Lemma Let C_p and C_q be cyclic groups of distinct prime orders p and q respectively. Let $G = C_p \times C_q$. Then there exists $i \in \Delta_Z(G)$ such that

$$\Delta_Z^{\omega}(G)\ (1-i) = 0.$$

Proof. Choose integers a and b such that $1 = pa+qb$. Let x and y be generators of C_p and C_q respectively. Then

$$\begin{aligned}(x-1)(y-1) &= (ap+bq)(x-1)(y-1)\\ &= ap(x-1)(y-1) + bq(x-1)(y-1)\\ &= -a\tbinom{p}{2}(x-1)^2(y-1) - \ldots - a(x-1)^p(y-1)\\ &\quad -b\tbinom{q}{2}(x-1)(y-1)^2 - \ldots - b(x-1)(y-1)^q\\ &= (x-1)(y-1)i \quad \text{with} \quad i \in \Delta_Z(G).\end{aligned}$$

Since $\Delta_Z^{\omega}(G) = (x-1)(y-1)Z(G)$, the Lemma is proved.

With the help of the above Lemma, we can easily finish the nilpotent case.

5.2 Lemma If G is a finite nilpotent group, then there exists an element $i \in \Delta_Z(G)$ such that

$$\Delta_Z^{\omega}(G)(1-i) = 0.$$

Proof. We proceed by induction on the order $o(G)$ of G, the case $o(G) = 1$ being trivial. Let G be a finite nilpotent group with $o(G) > 1$ and suppose the result holds for finite nilpotent groups of smaller order. If G is a prime power group, then $\Delta_Z^{\omega}(G) = 0$ and we are done. Suppose G is not a prime power group. Then we can find two elements x and y in the centre of G having distinct prime orders p and q. Let $A = \langle x \rangle$ and $B = \langle y \rangle$. By induction hypothesis, there exist $i_1, i_2 \in \Delta_Z(G)$ such that $\Delta_Z^{\omega}(G)\ (1-i_1) \subseteq \Delta_Z(G,A)$ and $\Delta_Z^{\omega}(G)(1-i_2) \subseteq \Delta_Z(G,B)$. But then $\Delta_Z^{\omega}(G)(1-i_1)(1-i_2) \subseteq \Delta_Z(G,A) \cap \Delta_Z(G,B)$ $= \Delta_Z(G,A)\Delta_Z(G,B)$ (Lemma 3.2) $= \Delta_Z^{\omega}(A\times B)Z(G)$. By Lemma 5.1, there exists $i_3 \in \Delta_Z(A\times B)$ such that $\Delta_Z^{\omega}(A\times B)(1-i_3) = 0$. Hence $\Delta_Z^{\omega}(G)(1-i_1)(1-i_2)(1-i_3) = 0$ which gives the desired result.

5.3 Theorem ([60],[91]). Let G be a finitely generated nilpotent group and R a commutative Noetherian ring with identity. Then there exists $i \in \Delta_R(G)$ such that

$$\Delta_R^{\omega}(G)(1-i) = 0.$$

Proof. Let S_{p_i}, $i = 1,2,\ldots,n$ be the Sylow subgroup of G. By Lemma 5.2, there exists an element $i_o \in \Delta_R(G)$ such that

$$\Delta_R(G,S_{p_j})\Delta_R(G,S_{p_k})(1-i_o) = 0$$

for all j,k, $1 \leq j, k \leq n$, $j \neq k$. Since R is a commutative Noetherian ring, by Krull's Theorem we can find $r_j \in p_jR$ such that $R^+(p_j)(1-r_j) = 0$ ([1], p. 110). Moreover, we can assume that $r_j \in p_j^{m_j}R$, where $p_j^{m_j}$ is the order of S_{p_j}, since $\bigcap_n p_j^n R = \bigcap_n p^{m_j n} R$. Let $r_j = p^{m_j} s_j$, $s_j \in R$. Then $R^+(p_j)\Delta_R(G,S_{p_j})(1-r_j+s_j \sum_{g\in S_{p_j}} g) = 0$. Let $t_j = r_j - s_j \sum_{g\in S_{p_j}} g$. Then $t_j \in \Delta_R(G)$ and is in the centre of $R(G)$. We construct t_j, $j = 1,2,\ldots,n$, for each p_j dividing the order of the torsion subgroup of G. Then, it is clear from Theorem 3.1 that $i = 1 - (1-i_o)\prod_{j=1}^{n}(1-t_j)$ satisfies

$$\Delta_R^{\omega}(G)(1-i) = 0.$$

The following result of Parmenter-Passi [58] gives the necessary and sufficient conditions for $\Delta_Z(G)$ to satisfy the intersection theorem.

5.4 Theorem Let G be a group. Then the following statements are equivalent:

(i) There exists an element $i \in \Delta_Z(G)$ such that $\Delta_Z^{\omega}(G)(1-i) = 0$

(ii) Either $\Delta_Z^{\omega}(G) = 0$,

or (a) the torsion elements form a non-identity finite nilpotent subgroup of G and

(b) if G has a p-element, then $G(p)$ is the set of torsion elements of G with order prime to p.

Proof. "(i) $\Rightarrow$ (ii)"

We first show that $D_{\omega,Z}(G) = 1$ and so G is residually nilpotent. Let $g \in D_{\omega,Z}(G)$. Then $(g-1)(1-i) = 0$. Therefore, g must be a periodic element and

$$1-i = (1+g+g^2+\ldots+g^{m-1})a$$

for some $a \in Z(G)$, where m is the order of g (see [75] Chapter III, Section 1 for augmentation annihilators). Applying the augmentation map $\varepsilon : Z(G) \to Z$ to both sides of the above equation, we get $1 = m\,\varepsilon(a)$. Hence $m = 1$ and so $D_{\omega,Z}(G) = 1$.

Suppose $\Delta_Z^{\omega}(G) \neq 0$. Then, by Theorem 2.30 of Chapter VI, $P(G) \neq 1$ and G is not discriminated by nilpotent p-groups of bounded exponents. Thus, in particular, $G(p) \neq 1$ for every prime p. Let $1 \neq h \in G(p)$. Then $(g-1)(h-1) \in \Delta_Z^{\omega}(G)$ for all $g \in G_{\omega p}$ (Theorem 2.3). Therefore, $(g-1)(h-1)(1-i) = 0$ for all $g \in G_{\omega p}$, while $(h-1)(1-i) \neq 0$ (for otherwise the argument given in the first paragraph above will imply that $h = 1$). It follows that $G_{\omega p}$ is finite for all primes p (because the right annihilator of $\Delta_Z(G_{\omega p}) \neq 0$.

Let $\tau(G) = \{x \in G \mid x$ has finite order$\}$. Suppose $\tau(G) = 1$, i.e. G is torsion-free. Since G is not discriminated by nilpotent p-groups of bounded exponents, we can choose non-identity elements $g_1, g_2, \ldots, g_n$ in G such that

$$(g-1)(g_1-1)(g_2-1)\ldots(g_n-1) \in \Delta_Z^{\omega}(G)$$

for all $g \in P(G)$ (see the proof of Theorem 2.30, Chapter VI). Therefore, $(g-1)(g_1-1)(g_2-1)\ldots(g_n-1)(1-i) = 0$ for all $g \in P(G)$. By annihilator considerations, this is impossible when G is torsion-free and $P(G) \neq 1$. Consequently, $\tau(G) \neq 1$ and so $G_{\omega p} \neq 1$ for some prime p.

Let p be a prime such that $G_{\omega p} \neq 1$. Let $1 \neq g \in G_{\omega p}$. Then $(g-1)(1-i) \neq 0$ and $(h-1)(g-1)(1-i) = 0$ for all $h \in G(p)$. Hence $G(p)$ is finite for all such primes p. It now follows that $\tau(G)$ is a finite nilpotent subgroup of G. For, let $x \in \tau(G)$. Then we can write $x = yz$, where y is a p-element and $z \in \tau(G)$ has order coprime to p. Thus $y \in G_{\omega p}$ and $z \in G(p)$. This implies that $\tau(G) \subseteq G_{\omega p}G(p)$. Since each of $G_{\omega p}$ and $G(p)$ is a finite normal subgroup of G, the assertion follows.

To complete the proof we show that $G(p) \cap G_{\omega p} = 1$ for all primes p. Let $1 \neq x \in G(p) \cap G_{\omega p}$. Then $(x-1)^2 \in \Delta_Z^{\omega}(G)$. Therefore, $(x-1)^2(1-i) = 0$. This implies $(x-1)(1-i) = 0$ which forces x to be 1. Hence $G_{\omega p} \cap G(p) = 1$. We thus conclude that whenever $G_{\omega p} \neq 1$, then $G(p) = \prod_{q \neq p} G_{\omega p}$. This completes the proof of the implication "(i) $\Rightarrow$ (ii)".

"(ii) $\Rightarrow$ (i)".

If $\Delta_Z^{\omega}(G) = 0$, then (i) holds trivially. Suppose both (a) and (b) hold. By Proposition 1.3, for every prime p, $G/G(p)$ is residually nilpotent p-group of bounded exponent. Therefore, by Theorem 2.11 of Chapter VI, $\Delta_Z^{\omega}(G/G(p)) = 0$. Hence $\Delta_Z^{\omega}(G) \subseteq \Delta_Z(G, G(p))$ for all primes p. It follows from Lemma 3.2, Theorem 2.5 and the hypotheses (a) and (b) that $\Delta_Z^{\omega}(G) = Z(G)\Delta_Z^{\omega}(\tau(G))$. The required implication therefore follows from Lemma 5.2.

CHAPTER VIII

THE ASSOCIATED GRADED RING OF A GROUP RING

The powers of the augmentation ideal provide a filtration of a group ring and we can define the associated graded ring. In this Chapter we study the structure of this associated graded ring (see also [69] for a survey on this topic).

1. THE ASSOCIATED GRADED RING OF A GROUP RING

Let G be a group and let $R(G)$ be its group ring over a commutative ring R with identity. The powers $\Delta_R^i(G)$ of the augmentation ideal $\Delta_R(G)$, $i \geq 0$, provide a filtration of $R(G)$. Consider the direct sum

$$\mathfrak{G}_R(G) = \sum_{i \geq 0} \Delta_R^i(G)/\Delta_R^{i+1}(G)$$

of R-modules $\Delta_R^i(G)/\Delta_R^{i+1}(G)$. We regard $\mathfrak{G}_R(G)$ as a graded R-module with the convention that the elements of $\Delta_R^i(G)/\Delta_R^{i+1}(G)$ are homogeneous of degree i. We define multiplication on $\mathfrak{G}_R(G)$ as follows. For $x_i \in \Delta_R^i(G)$, let $\underline{x}_i = x_i + \Delta_R^{i+1}(G)$. Then, for $\underline{x}_i \in \Delta_R^i(G)/\Delta_R^{i+1}(G)$, $\underline{x}_j \in \Delta_R^i(G)/\Delta_R^{j+1}(G)$, we define $\underline{x}_i\underline{x}_j = \underline{x_i x_j}$.

Note that $x_i x_j \in \Delta_R^{i+j}(G)$ and that the product is well-defined, i.e. $\underline{x_i x_j}$ depends only on the cosets of x_i and x_j mod $\Delta_R^{i+1}(G)$ and $\Delta_R^{j+1}(G)$ respectively. The product of two arbitrary elements of $\mathfrak{G}_R(G)$ is defined by extending the above product to all of $\mathfrak{G}_R(G)$ by linearity: if $\underline{x} = \Sigma \underline{x}_i$ and $\underline{y} = \Sigma \underline{y}_j$ are decomposition of $\underline{x}$ and $\underline{y} \in \mathfrak{G}_R(G)$ into their homogeneous components, then

$$\underline{x}\underline{y} = \sum_{i,j} \underline{x}_i \underline{y}_j .$$

With this multiplication $\mathfrak{G}_R(G)$ becomes a (associative) graded ring and we call it the <u>associated graded ring of</u> $R(G)$. Note that $\mathfrak{G}_R(G)$ is in fact a graded R-algebra.

If k is a field, then the structure of $\mathfrak{G}_k(G)$ is completely described by a theorem of Quillen [77] in terms of the universal envelope of a certain Lie ring arising from an N-series, namely the dimension series of G over k. We, therefore, describe a construction [38] which associates to an N-series a graded Lie ring.

2. THE GRADED LIE RING ASSOCIATED TO AN N-SERIES

Let $G = H_1 \supseteq H_2 \supseteq \dots \supseteq H_i \supseteq \dots$ be an N-series in a group G

(Chapter III, 1.1). Then H_i/H_{i+1} is an Abelian group for all $i \geq 1$ because $(H_i,H_i) \subseteq H_{2i} \subseteq H_{i+1}$. We write these groups additively. Thus, if $\tilde{x}_i$ denotes the element x_iH_{i+1}, $x_i \in H_i$, then

$$\tilde{x}_i + \tilde{x}_i' = \widetilde{x_i x_i'} \quad , \quad x_i, x_i' \in H_i .$$

Consider the direct sum

$$\mathfrak{L}(G) = \sum_{i \geq 1} H_i/H_{i+1} .$$

We give the Abelian group $\mathfrak{L}(G)$ a grading with the convention that the elements of H_i/H_{i+1} are homogeneous of degree i. The Abelian group $\mathfrak{L}(G)$ can be given the structure of a Lie ring as follows:

For $\tilde{x}_i \in H_i/H_{i+1}$, $\tilde{x}_j \in H_j/H_{j+1}$, define the product $[\tilde{x}_i,\tilde{x}_j]$ by setting

$$(2.1) \qquad [\tilde{x}_i,\tilde{x}_j] = (\widetilde{x_i,x_j}) \in H_{i+j}/H_{i+j+1} ,$$

where (x_i,x_j) is the group commutator $x_i^{-1}x_j^{-1}x_ix_j \in H_{i+j}$.

The commutator identities

$$(2.2) \qquad (x,yz) = (x,z)(x,y)^z \qquad (x,y,z \in G)$$

and

$$(2.3) \qquad (xy,z) = (x,z)^y(y,z) \qquad (x,y,z \in G) ,$$

where $u^v = v^{-1}uv$ $(u,v \in G)$ ensure that $[\tilde{x}_i,\tilde{x}_j]$ is well-defined. For, let $x_{j+1} \in H_{j+1}$, $x_{i+1} \in H_{i+1}$. Then

$$(x_i,x_{j+1}x_j) = (x_i,x_j)(x_i,x_{j+1})^{x_j} ,$$

$$(x_{i+1}x_i,x_j) = (x_{i+1},x_j)^{x_i}(x_i,x_j)$$

and $(x_i,x_{j+1})^{x_j}$, $(x_{i+1},x_j)^{x_i} \in H_{i+j+1}$.

Therefore, the element $(\widetilde{x_i,x_j})$ of H_{i+j}/H_{i+j+1} depends only on the cosets of x_i and x_j mod H_{i+1} and H_{j+1} respectively. The identities (2.2) and (2.3) further imply that the product $[\tilde{x}_i,\tilde{x}_j]$ satisfies the following identities:

$$[\tilde{x}_i,\tilde{x}_j+\tilde{x}_j'] = [\tilde{x}_i,\tilde{x}_j] + [\tilde{x}_i,\tilde{x}_j']$$

and

$$[\tilde{x}_i+\tilde{x}_i',\tilde{x}_j] = [\tilde{x}_i,\tilde{x}_j] + [\tilde{x}_i',\tilde{x}_j] .$$

The commutator identities

$$(x,x) = 1 \quad \text{and} \quad (x,y) = (y,x)^{-1}$$

imply that the product $[\tilde{x}_i,\tilde{x}_j]$ satisfies

$$[\tilde{x}_i,\tilde{x}_i] = 0 \quad\text{and}\quad [\tilde{x}_i,\tilde{x}_j] = -[\tilde{x}_j,\tilde{x}_i] .$$

We extend the definition of the product $[\tilde{x},\tilde{y}]$ to arbitrary elements of $\mathfrak{L}(G)$ by linearity. If $\tilde{x} = \Sigma\tilde{x}_i$ and $\tilde{y} = \Sigma\tilde{y}_j$ are the decompositions of $\tilde{x}$ and $\tilde{y}$ into their homogeneous components, then we define

$$(2.4) \qquad [\tilde{x},\tilde{y}] = \sum_{i,j} [\tilde{x}_i,\tilde{y}_j] .$$

Then the product $[\tilde{x},\tilde{y}]$ is a bilinear product on $\mathfrak{L}(G)$ which satisfies

$$(2.5) \qquad [\tilde{x},\tilde{x}] = 0 \quad \text{for all} \quad \tilde{x} \in \mathfrak{L}(G) .$$

We have the following commutator identity in G:

$$(2.6) \qquad ((x,y^{-1}),z)^y((y,z^{-1}),x)^z((z,x^{-1}),y)^x = 1 \qquad (x,y,z \in G) .$$

This identity implies that in $\mathfrak{L}(G)$, we have

$$(2.7) \qquad [[\tilde{x},\tilde{y}],\tilde{z}] + [[\tilde{y},\tilde{z}],\tilde{x}] + [[\tilde{z},\tilde{x}],\tilde{y}] = 0$$

for all $\tilde{x},\tilde{y},\tilde{z} \in \mathfrak{L}(G)$, i.e. the Jacobi identity is satisfied by the product $[\tilde{x},\tilde{y}]$ in $\mathfrak{L}(G)$. Hence $\mathfrak{L}(G)$ has the structure of a graded Lie ring with $[\tilde{x},\tilde{y}]$ as the Lie product. We call $\mathfrak{L}(G)$ the graded Lie ring associated to the N-series $\{H_i\}$.

We have seen in Chapter III that the dimension series $\{D_{i,R}(G)\}$ of a group G over any commutative ring R with identity is always an N-series. We denote by $\mathfrak{L}_R(G)$ the Lie ring $\sum_{i\geq 1} D_{i,R}(G)/D_{i+1,R}(G)$ associated to the N-series $\{D_{i,R}(G)\}$. Then we have a natural additive monomorphism

$$(2.8) \qquad \theta = \theta_R(G): \mathfrak{L}_R(G) \to \mathfrak{G}_R(G)$$

given by $\tilde{x}_i \mapsto x_i - 1 + \Delta_R^{i+1}(G)$ on the homogeneous elements of degree i. Clearly θ is a homomorphism of Lie rings when $\mathfrak{G}_R(G)$ is regarded as a Lie ring under commutation, i.e. $[\alpha,\beta] = \alpha\beta - \beta\alpha$, $\alpha,\beta \in \mathfrak{G}_R(G)$.

On tensoring $\mathfrak{L}_R(G)$ over Z with R we get a Lie algebra $R \otimes_Z \mathfrak{L}_R(G)$ over R and the homomorphism $\theta : \mathfrak{L}_R(G) \to \mathfrak{G}_R(G)$ induces a Lie algebra homomorphism

$$(2.9) \qquad \theta : R \otimes_Z \mathfrak{L}_R(G) \to \mathfrak{G}_R(G) .$$

Let us regard $R(G)$ as a Lie algebra over R by commutation. Then the Lie powers $\Delta_R^{[i]}(G)$ of the augmentation ideal $\Delta_R(G)$ are defined inductively as follows:

$$\Delta_R^{[1]}(G) = \Delta_R(G) , \quad \Delta_R^{[i]}(G) = \text{the R-submodule of } R(G) \text{ spanned}$$

by the elements $[\alpha,\beta] = \alpha\beta-\beta\alpha$, $\alpha \in \Delta_R(G)$, $\beta \in \Delta_R^{[i-1]}(G)$, $i \geq 2$.

The following Proposition is easily proved by induction on i .

2.10 Proposition. $\Delta_R^{[i]}(G) + \Delta_R^{i+1}(G) = \Delta_R(G,\gamma_i(G)) + \Delta_R^{i+1}(G)$ for all $i \geq 1$.

Let us examine the Lie ring $\mathfrak{L}_R(G)$ when R is the field $\mathbb{Q}$ of rational numbers or F_p , the field of p elements.

3. THE LIE RING $\mathfrak{L}_{\mathbb{Q}}(G)$

3.1 Theorem. There is a Lie algebra isomorphism

$$\theta : \mathbb{Q} \otimes_Z \mathfrak{L}_{\mathbb{Q}}(G) \to \sum_{i\geq 1} \Delta_{\mathbb{Q}}^{[i]}(G) + \Delta_{\mathbb{Q}}^{i+1}(G)/\Delta_{\mathbb{Q}}^{i+1}(G)$$

between $\mathbb{Q} \otimes_Z \mathfrak{L}_{\mathbb{Q}}(G)$ and the sub-Lie algebra of $\mathfrak{G}_{\mathbb{Q}}(G)$ generated by the homogeneous elements of degree 1.

Proof. Recall that the dimension subgroups of a group G over $\mathbb{Q}$ are given by $D_{i,\mathbb{Q}}(G) = \sqrt{\gamma_i(G)}$ (Chapter IV, Theorem 1.5).
The homomorphism

(2.9) $\theta : \mathbb{Q} \otimes_Z \mathfrak{L}_{\mathbb{Q}}(G) \to \mathfrak{G}_{\mathbb{Q}}(G)$ is a monomorphism (of Lie algebras). The image of $\mathbb{Q} \otimes_Z D_{i,\mathbb{Q}}(G)/D_{i+1,\mathbb{Q}}(G)$ under θ (by definitions) is equal to $\Delta_{\mathbb{Q}}(G,D_{i,\mathbb{Q}}(G)) + \Delta_{\mathbb{Q}}^{i+1}(G)/\Delta_{\mathbb{Q}}^{i+1}(G)$. It is easy to check that for all $i \geq 1$ $\Delta_{\mathbb{Q}}(G,\gamma_i(G)) + \Delta_{\mathbb{Q}}^{i+1}(G) = \Delta_{\mathbb{Q}}(G,D_{i,\mathbb{Q}}(G)) + \Delta_{\mathbb{Q}}^{i+1}(G)$. Thus the result follows by using Proposition 2.10.

4. THE LIE ALGEBRA $\mathfrak{L}_{F_p}(G)$

The dimension subgroups $D_{n,F_p}(G)$ of a group G over the field F_p of p elements are given by

$$D_{n,F_p}(G) = \prod_{ip^j\geq n} \gamma_i(G)^{p^j} , \quad n\geq 1 ,$$

and the series $\{D_{n,F_p}(G)\}_{n\geq 1}$ is a restricted N-series relative to p (Chapter IV, Theorem 1.9 and Chapter III, Corollary 1.3).

Since $D_{i,F_p}(G)/D_{i+1,F_p}(G)$ is an elementary Abelian p-group for all $i \geq 1$, $\mathfrak{L}_{F_p}(G)$ is a Lie algebra over F_p . In fact, $\mathfrak{L}_{F_p}(G)$ is a restricted Lie algebra over F_p . To see this let us recall the following

4.1 Definition [34]. A restricted Lie algebra L over a field k of characteristic $p > 0$ is a Lie algebra over k in which there is defined a map $a \mapsto a^{[p]}$ such that

(i) $(\alpha a)^{[p]} = \alpha^p a^{[p]}$, $a \in L$, $\alpha \in k$,

(ii) $(a+b)^{[p]} = a^{[p]} + b^{[p]} + \sum_{i=1}^{p-1} s_i(a,b)$,

$a,b \in L$, where $is_i(a,b)$ is the coefficient of λ^{i-1} in

$$[\ldots[[a,a\lambda+b],a\lambda+b],\ldots,a\lambda+b] \text{ ; } \quad (p-1 \text{ times})$$

(iii) $[a,b^{[p]}] = \underbrace{[\ldots[[a,b],b],\ldots,b]}_{p \text{ times}}$

$a,b \in L$.

If A is any associative algebra over a field k of characteristic $p > 0$, then A is a restricted Lie algebra with $[a,b] = ab-ba$ and $a^{[p]} = a^p$ [34]. Thus the associated graded ring $\mathfrak{G}_{F_p}(G)$ carries the structure of a restricted Lie algebra over F_p. Now we have a monomorphism (2.8) of Lie algebras

$$\theta : \mathfrak{L}_{F_p}(G) \to \mathfrak{G}_{F_p}(G)$$

which has the property that its image is closed under the map $a \to a^{[p]}$. For, if $\tilde{x}_i = x_i D_{i+1,F_p}(G)$, $x_i \in D_{i,F_p}(G)$ is a homogeneous element of degree i, then $\theta(\tilde{x}_i)^{[p]} = (x_i - 1 + \Delta_{F_p}^{i+1}(G))^p = (x_i^p - 1) + \Delta_{F_p}^{ip+1}(G) = \theta(\tilde{x}_i^p)$ (note that $x_i^p \in D_{ip,F_p}(G)$). Further, if a and b are in the image of θ, then $s_i(a,b)$, $1 \leq i \leq p-1$, being in the sub-Lie algebra generated by a and b, are all in the image. Hence the assertion about the image of θ holds. Consequently, θ induces on $\mathfrak{L}_{F_p}(G)$ the structure of a restricted Lie algebra: $\tilde{x}_i^{[p]} = \tilde{x}_i^p$. [It has been shown by Lazard ([38], Corollary 6.8) that if $\{H_i\}$ is any restricted N-series relative to a prime p, then the associated Lie ring $\sum_{i \geq 1} H_i/H_{i+1}$ is a restricted Lie algebra over F_p with $\tilde{x}_i^{[p]} = \tilde{x}_i^p$.] Moreover, if k is any field of characteristic $p > 0$, then $\mathfrak{L}_{F_p}(G)$ induces on $k \otimes_Z \mathfrak{L}_k(G)$ $(= k \otimes_Z \mathfrak{L}_{F_p}(G))$ the structure of a restricted Lie algebra over k.

For a subset S of $F_p(G)$, let S^{p^j} denote the subspace of $F_p(G)$ generated by the elements s^{p^j}, $s \in S$.

4.2 Theorem. There is a monomorphism

$$\theta : \mathfrak{L}_{F_p}(G) \to \mathfrak{G}_{F_p}(G)$$

of restricted Lie algebras such that

$$\mathrm{Im}(\theta) = \sum_{n\geq 1} L_{n,p}(G)/\Delta_{F_p}^{n+1}(G) = \text{the restricted sub-Lie}$$

algebra of $\mathfrak{G}_{F_p}(G)$ generated by its homogeneous elements of degree 1, where

$$L_{n,p}(G) = \Sigma_{ip^j \geq n} (\Delta_{F_p}^{[i]}(G))^{\underline{p}^j} + \Delta_{F_p}^{n+1}(G) .$$

Proof. Since $D_{n,F_p}(G) = \prod_{ip^j \geq n} \gamma_i(G)^{p^j}$, it follows, from definitions, that as a restricted Lie algebra over F_p, $\mathfrak{L}_{F_p}(G)$ is generated by $G/D_{2,F_p}(G)$. Therefore, $\mathrm{Im}(\theta)$ is generated, as a restricted Lie algebra, by $\theta(G/D_{2,F_p}(G)) = \Delta_{F_p}(G)/\Delta_{F_p}^2(G)$ (see Chapter II, Example 1.5). The assertion about the description of $\mathrm{Im}(\theta)$ as $\sum_{n\geq 1} L_{n,p}(G)/\Delta_{F_p}^{n+1}(G)$ follows easily from Proposition 2.10 [70].

5. THE ASSOCIATED GRADED ALGEBRA OF A GROUP ALGEBRA

Let R be a commutative ring with identity. Let M be an R-module. We denote by $\mathfrak{T}(M)$ the tensor algebra of M over R. $\mathfrak{T}(M) = \sum_{i\geq 0} T^i(M)$, where $T^0(M) = R$ and $T^i(M) = \underbrace{M \otimes M \otimes \dots \otimes M}_{i \text{ times}}$ (tensor product over R) for $i \geq 1$.

Let L be a Lie algebra over R. We denote by $\mathfrak{U}(L)$ the universal envelope of L over R: $\mathfrak{U}(L)$ = the quotient algebra of $\mathfrak{T}(L)$ modulo the 2-sided ideal generated by all elements of the type $a\otimes b - b\otimes a - [a,b]$ $(a,b \in L)$.

Let L be a restricted Lie algebra over a field k of characteristic $p > 0$. We denote by $\mathfrak{U}^r(L)$ the restricted universal envelope of L over k: $\mathfrak{U}^r(L)$ = the quotient algebra of $\mathfrak{T}(L)$ modulo the 2-sided ideal generated by all elements of the type $a\otimes b - b\otimes a - [a,b]$ and $a^{[p]} - \underbrace{a\otimes a\otimes \dots \otimes a}_{p \text{ times}}$ $(a,b \in L)$.

Let L be a Lie algebra over a field k of characteristic zero, then the natural map $L \to \mathfrak{U}(L)$ is an embedding of the Lie algebra L into $\mathfrak{U}(L)$ regarded as a Lie algebra under commutation. In this case the Poincaré-Birkhoff-Witt Theorem states that if x_i, $i \in I$, is a k-basis of L, and I is linearly ordered, then the monomials $x_{i_1}^{\alpha_1} x_{i_2}^{\alpha_2} \dots x_{i_n}^{\alpha_n}$ with $i_1 < i_2 < \dots < i_n$ form a basis of $\mathfrak{U}(L)$. A similar result holds in the case of restricted Lie algebras over a field of characteristic $p > 0$. In this case a basis for the restricted universal envelope is given by the monomials $x_{i_1}^{\alpha_1} x_{i_2}^{\alpha_2} \dots x_{i_n}^{\alpha_n}$ with $i_1 < i_2 < \dots < i_n$ and $0 \leq \alpha_i \leq p-1$.

A connection between the universal envelopes of Lie algebras and the associated graded ring $\mathfrak{G}_R(G)$ of a group ring $R(G)$ is provided by the observation that the homomorphism (2.9)

$$\theta : R \otimes_Z \mathfrak{L}_R(G) \to \mathfrak{G}_R(G)$$

induces a homomorphism of graded R-algebras

$$(5.1) \qquad \theta = \theta_R(G) : \mathfrak{U}(R \otimes \mathfrak{L}_R(G)) \to \mathfrak{G}_R(G)$$

which is in fact an epimorphism (the grading on $\mathfrak{U}(R \otimes_Z \mathfrak{L}_R(G))$ is the one induced by the grading on $\mathfrak{L}_R(G)$). If R is a field of characteristic $p > 0$, then in (5.1) we can take $\mathfrak{U}^r$ instead of $\mathfrak{U}$.

5.2 <u>Theorem</u> [77]. <u>Let</u> G <u>be a group</u>.

(i) <u>If</u> k <u>is a field of characteristic zero</u>, <u>then</u>

$$\theta : \mathfrak{U}(k \otimes_Z \mathfrak{L}_k(G)) \cong \mathfrak{G}_k(G) .$$

(ii) <u>If</u> k <u>is a field of characteristic</u> $p > 0$, <u>then</u>

$$\theta : \mathfrak{U}^r(k \otimes_Z \mathfrak{L}_k(G)) \cong \mathfrak{G}_k(G)$$

[Quillen [77] proves this result by using the structure theorems of Milnor-Moore [54] on Hopf algebras. We present an alternative proof based on the work of Chapters III and IV.]

We need some observations about the universal envelope $\mathfrak{U}(R \otimes_Z \mathfrak{L}_R(G))$.

Let $f : G \to H$ be a homomorphism. Then $f(D_{i,R}(G)) \subseteq D_{i,R}(H)$ for all $i \geq 1$ and we can define a homomorphism of Lie rings $\mathfrak{L}_R(G) \to \mathfrak{L}_R(H)$, $xD_{i+1,R}(G) \mapsto f(x)D_{i+1,R}(H)$, $x \in D_{i,R}(G)$. This homomorphism of Lie rings in turn induces a homomorphism (of graded R-algebras)

$$(5.3) \qquad \mathfrak{U}(f) : \mathfrak{U}(R \otimes_Z \mathfrak{L}_R(G)) \to \mathfrak{U}(R \otimes_Z \mathfrak{L}_R(H)) .$$

Let $I(G,R)$ be the two sided ideal of the tensor algebra $\mathfrak{T}(R \otimes_Z \mathfrak{L}_R(G))$ generated by all elements of the type $x \otimes y - y \otimes x - [x,y]$

$(x,y \in R \otimes_Z \mathfrak{L}_R(G))$. Then a typical homogeneous element of $\mathfrak{U}(R \otimes_Z \mathfrak{L}_R(G))$ is a finite sum of elements of the type

$$r \otimes \tilde{x}_{i_1} \otimes \tilde{x}_{i_2} \otimes \dots \otimes \tilde{x}_{i_r} + I(G,R)$$

where $r \in R$, $\tilde{x}_{i_j} = x_{i_j} D_{i_j+1,R}(G)$, $x_{i_j} \in D_{i_j,R}(G)$, (see Section 2) $\Sigma i_j = n$, n being the degree of the element. The homomorphism $\mathfrak{U}(f)$ maps $r \otimes \tilde{x}_{i_1} \otimes \tilde{x}_{i_2} \otimes \dots \otimes \tilde{x}_{i_r} + I(G,R)$ into $r \otimes f(x_{i_1}) \otimes f(x_{i_2}) \otimes \dots \otimes f(x_{i_r}) + I(H,R)$. Under the homomorphism $\theta : \mathfrak{U}(R \otimes_Z \mathfrak{L}_R(G)) \to \mathfrak{G}_R(G)$, the element $r \otimes \tilde{x}_{i_1} \otimes \tilde{x}_{i_2} \otimes \dots \otimes \tilde{x}_{i_r} + I(G,R)$ is mapped into $r(x_{i_1}-1)(x_{i_2}-1)\dots(x_{i_r}-1) + \Delta_R^{n+1}(G)$.

Let $\mathfrak{U}_n(R \otimes_Z \mathfrak{L}_R(G))$ denote the homogeneous component of degree n of $\mathfrak{U}(R \otimes_Z \mathfrak{L}_R(G))$. Let $z = \Sigma\, r \otimes \tilde{x}_{i_1} \otimes \tilde{x}_{i_2} \otimes \dots \otimes \tilde{x}_{i_r} + I(G,R)$ be an element of $\mathfrak{U}_n(R \otimes_Z \mathfrak{L}_R(G))$. Suppose $\theta(z) = 0$. then the element $r(x_{i_1}-1)(x_{i_2}-1)\dots(x_{i_r}-1)$ lies in $\Delta_R^{n+1}(G)$. Let H be the (finitely generated) subgroup of G generated by all elements of G which occur in the sum $\Sigma\, r(x_{i_1}-1)(x_{i_2}-1)\dots(x_{i_r}-1)$ and in its expression as an element of $\Delta_R^{n+1}(G)$. Now in $\mathfrak{U}_n(R \otimes_Z \mathfrak{L}_R(H))$ we have the element $w = \Sigma\, r \otimes \tilde{x}_{i_1} \otimes \tilde{x}_{i_2} \otimes \dots \otimes \tilde{x}_{i_r} + I(H,R)$ which has the properties that

(a) $\mathfrak{U}(i)(w) = z$, where $\mathfrak{U}(i)$ is the homomorphism induced by the inclusion $i : H \to G$;

(b) $\theta_R(H)(w) = 0$.

Thus, if for every finitely generated subgroup H of G

$$\theta : \mathfrak{U}(R \otimes_Z \mathfrak{L}_R(H)) \to \mathfrak{G}_R(H)$$

is an isomorphism, then

$$\theta : \mathfrak{U}(R \otimes_Z \mathfrak{L}_R(G)) \to \mathfrak{G}_R(G)$$

is also an isomorphism. Another observation which we need is that if $H = G/D_{n+1,R}(G)$, then the restriction of $\mathfrak{U}(f)$, where $f : G \to H$ is the natural projection, to $\mathfrak{U}_i(R \otimes_Z \mathfrak{L}_R(G))$ is an isomorphism for $0 \leq i \leq n$.

Every homomorphism $f : G \to H$ induces a homomorphism of graded R-algebras

(5.4) $$\mathfrak{G}_R(f) : \mathfrak{G}_R(G) \to \mathfrak{G}_R(H)$$

$$(x_1-1)(x_2-1)\dots(x_n-1) + \Delta_R^{n+1}(G) \longmapsto (f(x_1)-1)(f(x_2)-1)\dots(f(x_n)-1) + \Delta_R^{n+1}(H) .$$

Clearly the diagram

$$\begin{array}{ccc} \mathfrak{U}(R\otimes_Z \mathfrak{L}_R(G)) & \xrightarrow{\mathfrak{U}(f)} & \mathfrak{U}(R\otimes_Z \mathfrak{L}_R(H)) \\ \downarrow \theta & & \downarrow \theta \\ \mathfrak{G}_R(G) & \xrightarrow{\mathfrak{G}_R(f)} & \mathfrak{G}_R(H) \end{array}$$

is commutative. As for the universal envelopes, if $H = G/D_{n+1,R}(G)$ then the restriction of $\mathfrak{G}_R(f)$, where $f : G \to H$ is the natural projection, to $\Delta_R^i(G)/\Delta_R^{i+1}(G)$ is an isomorphism for $0 \leqslant i \leqslant n$.

5.5 Proof of Theorem 5.2.

(i) Let k be a field of characteristic zero. The foregoing observations ensure that it is enough to prove the Theorem for finitely generated torsion-free nilpotent groups (recall that $D_{n,k}(G) = \sqrt{\gamma_n(G)}$). Let G be a finitely generated torsion-free nilpotent group. Then $D_{i,k}(G)/D_{i+1,k}(G)$ is a free Abelian group of finite rank for all $i \geqslant 1$. Let e_i = rank of $D_{i,k}(G)/D_{i+1,k}(G)$. By the Poincaró-Birkhoff-Witt Theorem we can compute the dimensions over k of the various homogeneous components $\mathfrak{U}_t(k\otimes_Z \mathfrak{L}_k(G))$ $(t \geqslant 0)$ in terms of the e_i's. On the other hand the dimensions of the homogeneous components $\Delta_k^t(G)/\Delta_k^{t+1}(G)$ can be computed from the results of Chapters III and IV (use Theorems 2.15 and 2.28 from Chapter III and Lemma 1.4 from Chapter IV). An easy comparison shows that the dimension of $\mathfrak{U}_t(k\otimes_Z \mathfrak{L}_k(G))$ = dimension of $\Delta_k^t(G)/\Delta_k^{t+1}(G)$ for all $t \geqslant 0$. Hence the epimorphism $\theta : \mathfrak{U}(k\otimes_Z \mathfrak{L}_k(G)) \to \mathfrak{G}_k(G)$ must be an isomorphism.

(ii) The proof is similar to that of (i) and is omitted.

A precise count of the dimensions gives the following results.

5.6 Theorem. Let G be a finitely generated nilpotent group of class c and k a field of characteristic zero. Let

$$f_t = \dim \Delta_k^t(G)/\Delta_k^{t+1}(G) \;, \quad e_t = \text{rank of } D_{t,k}(G)/D_{t+1,k}(G) \;.$$

Then a generating function $\sum_{t=0}^{\infty} f_t\zeta^t$ is given by

$$\sum_{t=0}^{\infty} f_t\zeta^t = (1-\zeta)^{-e_1}(1-\zeta^2)^{-e_2}\dots(1-\zeta^c)^{-e_c}$$

5.7 Theorem. Let k be a field of characteristic $p > 0$ and let G be a finite p-group. Let

$$\text{order of } D_{t,k}(G)/D_{t+1,k}(G) = p^{e_t}$$

and dimension of $\Delta_k^t(G)/\Delta_k^{t+1}(G) = f_t$.

Then a generating function $\sum_{t=o}^{\infty} f_t \zeta^t$

is given by $\sum_{t=o}^{\infty} f_t \zeta^t = \left(\frac{\zeta^p - 1}{\zeta - 1}\right)^{e_1} \left(\frac{\zeta^{2p} - 1}{\zeta - 1}\right)^{e_2} \cdots \left(\frac{\zeta^{dp} - 1}{\zeta - 1}\right)^{e_d}$

where d is the least integer for which $D_{d+1,k}(G) = 1$.

For more details on Theorems 5.6 and 5.7 see [75], Chapter III.

6. THE ASSOCIATED GRADED RING OF AN INTEGRAL GROUP RING

For a group G , let

$$LG = \sum_{i \geq 1} \gamma_i(G)/\gamma_{i+1}(G)$$

be the Lie ring associated to the lower central series of G (Section 2). Then the inclusion $\gamma_i(G) \subseteq D_{i,Z}(G)$ induces a Lie ring homomorphism

$$j : LG \to \mathfrak{L}_Z(G), \; x_i\gamma_{i+1}(G) \to x_i D_{i+1,Z}(G), \; x_i \in \gamma_i(G) .$$

The homomorphism j induces a homomorphism

$$\alpha : \mathfrak{U}(LG) \to \mathfrak{U}(\mathfrak{L}_Z(G))$$

between the universal envelopes of LG and $\mathfrak{L}_Z(G)$ (here LG and $\mathfrak{L}_Z(G)$ are regarded as Z-algebras (see Section 4)). Combined with the homomorphism (5.1)

$$\theta : \mathfrak{U}(\mathfrak{L}_Z(G)) \to \mathfrak{G}_Z(G)$$

we get a homomorphism

$$\alpha : \mathfrak{U}(LG) \to \mathfrak{G}_Z(G)$$

which is easily seen to be an epimorphism.

As in Section 5, let $\mathfrak{T}(M)$ denote the tensor algebra of an Abelian group (i.e. Z-module) M . The natural maps on homogeneous elements of degree 1 induce homomorphisms

$$\beta : \mathfrak{T}(G/\gamma_2(G)) \to \mathfrak{G}_Z(G) \; , \; \tilde{x} \mapsto x-1 + \Delta_Z^2(G) \; , \; x \in G \; ,$$

and

$$\gamma : \mathfrak{T}(G/\gamma_2(G)) \to \mathfrak{U}(LG) \; , \; \tilde{x} \mapsto \text{coset of } \tilde{x} \; , \; x \in G \; ,$$

both of which are onto and the diagram

$$(6.1) \qquad \begin{array}{ccc} \mathfrak{T}(G/\gamma_2(G)) & \xrightarrow{\gamma} & \mathfrak{U}(LG) \\ & {\scriptstyle\beta}\searrow \quad \swarrow{\scriptstyle\alpha} & \\ & \mathfrak{G}_Z(G) & \end{array}$$

is commutative.

Let us first consider free groups. If F is a free group, then by Magnus' Theorem $D_{n,Z}(F) = \gamma_n(F)$ for all $n \geq 1$ (Chapter IV, Corollary 3.2). Thus $LF = \mathfrak{L}_Z(F)$. For the associated graded ring of $Z(F)$ we have the following well-known result.

6.2 Theorem. If F is a free group, then

$$\mathfrak{B}(F/\gamma_2(F)) \cong \mathfrak{U}(LF) \cong \mathfrak{G}_Z(F) .$$

Proof. Let $(x_i)_{i \in I}$ be a basis of a free group F . It is well-known that $\Delta_Z(F)$ is a free left $Z(F)$-module with the elements $x_i - 1$, $i \in I$ as a basis ([30], p. 196). It follows from this that $\Delta_Z^n(F)$ $(n \geq 1)$ is a free left $Z(F)$-module with the elements $(x_{i_1}-1)(x_{i_2}-1)\ldots$ $\ldots(x_{i_n}-1)$, $i_j \in I$, as a basis. Consequently, for all $n \geq 1$, $\Delta_Z^n(F)/\Delta_Z^{n+1}(F)$ is a free Abelian group with the elements $(x_{i_1}-1)(x_{i_2}-1)\ldots(x_{i_n}-1) + \Delta_Z^{n+1}(F)$ as a basis. Let $\tilde{x}_i = x_i\gamma_2(F)$, $i \in I$. Then the elements $\tilde{x}_i$, $i \in I$ form a basis of $F/\gamma_2(F)$ as a free Abelian group. Therefore, $\underbrace{F/\gamma_2(F) \otimes F/\gamma_2(F) \otimes\ldots\otimes F/\gamma_2(F)}_{n \text{ times}}$ is a free Abelian group with the elements $\tilde{x}_{i_1} \otimes \tilde{x}_{i_2} \otimes\ldots\otimes \tilde{x}_{i_n}$, $i_j \in I$, as a basis. As

$$\beta(\tilde{x}_{i_1} \otimes \tilde{x}_{i_2} \otimes\ldots\otimes \tilde{x}_{i_n}) = (x_{i_1}-1)(x_{i_2}-1)\ldots(x_{i_n}-1) + \Delta_Z^{n+1}(F)$$

it follows that β is an isomorphism. That α is an isomorphism follows from the commutativity of (6.1) and the fact that γ is an epimorphism. Hence the Theorem is proved.

For any group G with all lower central factors free Abelian, Tahara (see [84]) has expressed $\mathfrak{G}_Z(G)$ as $\bigoplus_n W_n(G)$, where $W_n(G) = \bigoplus \bigotimes_i SP^{a_i}(\gamma_i(G)/\gamma_{i+1}(G))$, the sum is taken over all sequences of non-negative integers a_i such that $\Sigma i a_i = n$.

For an arbitrary group ring $R(G)$ we do not have a result similar to that of Quillen over fields. Stallings [92] has recently given a method of computing $\mathfrak{G}_R(G)$ using spectral sequence techniques. However, even the problem of precisely describing $\mathfrak{G}_Z(G)$ when G is finitely generated Abelian is quite complicated and is not completely settled. We, therefore, propose to restrict to the integral coefficients and make a term by term analysis of the homogeneous components

$Q_n(G) = \Delta_Z^n(G)/\Delta_Z^{n+1}(G)$ $(n \geq 0)$. The study of $Q_n(G)$ is naturally linked up with that of $P_n(G) = \Delta_Z(G)/\Delta_Z^{n+1}(G)$ because of the natural short exact sequence

$$0 \to Q_n(G) \to P_n(G) \to P_{n-1}(G) \to 0 .$$

(For a method of computing the additive structure of $P_n(G)$ in terms of a presentation of G , see [85]).

We first prove two reduction theorems. Recall that the <u>exponent</u> of a group G is the smallest integer $r > 0$ such that $x^r = 1$ for all $x \in G$ if it exists and infinite otherwise.

6.3 <u>Theorem</u> [62]. <u>If</u> G <u>and</u> H <u>are two groups of finite coprime exponents then</u>

$$P_n(G \oplus H) \cong P_n(G) \oplus P_n(H)$$

<u>and</u>

$$Q_n(G \oplus H) \cong Q_n(G) \oplus Q_n(H)$$

<u>for all</u> $n \geq 1$.

<u>Proof</u>. Consider the projections $f_1 \colon G \oplus H \to G$, $f_2 \colon G \oplus H \to H$ given by $f_1(xy) = x$, $f_2(xy) = y$, $x \in G$, $y \in H$. These homomorphisms induce in a natural way, homomorphisms

$$P_n(f_1) \; : P_n(G \oplus H) \to P_n(G)$$

and

$$P_n(f_2) \; : P_n(G \oplus H) \to P_n(H) .$$

Consider the homomorphism

$$\alpha : P_n(G \oplus H) \to P_n(G) \oplus P_n(H)$$

given by

$$\alpha(z) = P_n(f_1)(z) + P_n(f_2)(z) , \; z \in P_n(G \oplus H) .$$

The hypothesis on the exponents of G and H implies that $(x-1)(y-1) \in \Delta_Z^i(G \oplus H)$ for all $i \geq 1$, $x \in G$, $y \in H$ (see Chapter VI, proof of Theorem 1.2). From this observation it is easy to deduce that the homomorphism α is an isomorphism. The restriction of α to $Q_n(G \oplus H)$ gives an isomorphism

$$Q_n(G \oplus H) \cong Q_n(G) \oplus Q_n(H) .$$

6.4 <u>Theorem</u> [62]. <u>Let</u> $M = G \oplus Z$, <u>where</u> G <u>is an arbitrary group and</u> Z <u>is an infinite cyclic group</u>. <u>Then</u>

$$P_n(G \oplus Z) \cong P_n(G) \oplus P_{n-1}(G) \oplus \dots \oplus P_1(G) \oplus P_n(Z)$$

<u>and</u>

$$Q_n(G \oplus Z) \cong Q_n(G) \oplus Q_{n-1}(G) \oplus \dots \oplus Q_1(G) \oplus Q_n(Z)$$

for all $n \geq 1$.

Proof. Let x be a generator of the infinite cyclic group Z . Then $\Delta_Z(M) = \Delta_Z(G) + Z(M)(1-x)$. By induction on n (≥ 1) we have

$$\Delta_Z(M) = \Delta_Z(G) + Z(G)(1-x) + Z(G)(1-x)^2 + \ldots + Z(G)(1-x)^{n-1} + Z(M)(1-x)^n$$

and

$$\Delta_Z^n(M) = \Delta_Z^n(G) + \Delta_Z^{n-1}(G)(1-x) + \Delta_Z^{n-2}(G)(1-x)^2 + \ldots + \Delta_Z(G)(1-x)^{n-1} + Z(M)(1-x)^n .$$

It is easy to check that

$$\{Z(G) + Z(G)(1-x) + ZG(1-x)^2 + \ldots + Z(G)(1-x)^n\} \cap Z(M)(1-x)^{n+1} = 0$$

for all $n \geq 0$ (regard $Z(M)$ as a group ring of Z over $Z(G)$) .

We can therefore conclude that

$$P_n(M) \cong P_n(G) \oplus Z(G)/\Delta_Z^n(G) \oplus Z(G)/\Delta_Z^{n-1}(G) \oplus \ldots \oplus Z(G)/\Delta_Z(G)$$

and

$$Q_n(M) \cong Q_n(G) \oplus Q_{n-1}(G) \oplus Q_{n-2}(G) \oplus \ldots \oplus Q_1(G) \oplus Z(G)/\Delta_Z(G) \quad \text{for all } n \geq 1.$$

As $Z(G)/\Delta_Z^{m+1}(G) \cong P_m(G) \oplus Z$ for all $m \geq 1$, we have

$$P_n(M) \cong P_n(G) \oplus P_{n-1}(G) \oplus \ldots \oplus P_1(G) \oplus \underbrace{Z \oplus Z \oplus \ldots \oplus Z}_{n \text{ copies}} \quad \text{and}$$

$$Q_n(M) \cong Q_n(G) \oplus Q_{n-1}(G) \oplus \ldots \oplus Q_1(G) \oplus Z .$$

Taking G to be the trivial group, we get $P_n(Z) \cong \underbrace{Z \oplus Z \oplus \ldots \oplus Z}_{n \text{ copies}}$, $Q_n(Z) \cong Z$ and the Theorem is proved.

If we want to study the associated graded ring of a finitely generated Abelian group, then Theorems 6.3 and 6.4 imply that we can restrict ourselves to considering finite Abelian p-groups.

For an Abelian group G , $D_{2,Z}(G) = 1$, and, therefore, the Lie ring $\mathfrak{L}_Z(G)$ has all its terms of degree ≥ 2 equal to zero. The universal envelope $\mathfrak{U}(\mathfrak{L}_Z(G))$ in this case is the symmetric algebra

$$(6.5) \qquad \mathfrak{S}(G) = \sum_{n \geq 0} SP^n(G)$$

of the Abelian group G (see Chapter V, Section 1 for the symmetric powers $SP^n(G)$) and we have the epimorphism (5.1)

$$(6.6) \qquad \theta : \mathfrak{S}(G) \to \mathfrak{G}_Z(G)$$

which is defined on the n-th homogeneous component by

$$\theta(x_1 \hat{\otimes} x_2 \hat{\otimes} \ldots \hat{\otimes} x_n) = (x_1-1)(x_2-1)\ldots(x_n-1) + \Delta_Z^{n+1}(G) .$$

For finitely generated Abelian groups, $\mathfrak{S}(G)$ and $\mathfrak{G}_Z(G)$ can be interpreted as quotients of polynomial rings $Z[X_1, X_2, \ldots, X_k]$.

Let $G = Z_{n_1} \oplus Z_{n_2} \oplus \dots \oplus Z_{n_k}$, where Z_r denotes a cyclic group of order r if $r \geq 1$, $Z_o = Z$, the infinite cyclic group and n_i's are integers ≥ 0 . Let us write

(i) $f_i = \sum_{1 \leq s \leq n_i} \binom{n_i}{s} X_i^s$, (an element of the polynomial ring $Z[X_1, X_2, \dots, X_k]$, $f_i = 0$ if $n_i = 0$) ;

(ii) $\mathfrak{b}(G)$ = the ideal of $Z[X_1, X_2, \dots, X_k]$ generated by the elements f_i $(1 \leq i \leq k)$;

(iii) $\mathfrak{a}(G)$ = the ideal of $Z[X_1, X_2, \dots, X_k]$ generated by the leading forms of the elements of $\mathfrak{b}(G)$ (<u>leading form of a polynomial</u> is its non-zero homogeneous term of lowest degree);

(iv) $\mathfrak{s}(G)$ = the ideal of $Z[X_1, X_2, \dots, X_k]$ generated by the elements $n_i X_i$ $(1 \leq i \leq k)$.

6.7 <u>Proposition</u> [74]. <u>Let</u> $G = Z_{n_1} \oplus Z_{n_2} \oplus \dots \oplus Z_{n_k}$. <u>Then</u>

(i) $\mathfrak{S}(G) \cong Z[X_1, X_2, \dots, X_k]/\mathfrak{s}(G)$;

(ii) $\mathfrak{G}_Z(G) \cong Z[X_1, X_2, \dots, X_k]/\mathfrak{a}(G)$.

<u>Proof</u>. Let $a_1, a_2, \dots, a_k$ be a set of generators of the Abelian group G , where a_i is of order n_i if $n_i > 0$ and infinite otherwise.

(i) Consider the homomorphism $\alpha : Z[X_1, X_2, \dots, X_k] \to \mathfrak{S}(G)$ given by $X_i \mapsto a_i$ $(1 \leq i \leq k)$. Clearly its kernel contains the ideal $\mathfrak{s}(G)$. Since the map $a_i \mapsto X_i + \mathfrak{s}(G)$ can be extended to a homomorphism $\mathfrak{S}(G) \to Z[X_1, X_2, \dots, X_k]/\mathfrak{s}(G)$, α induces an isomorphism $Z[X_1, X_2, \dots, X_k]/\mathfrak{s}(G) \to \mathfrak{S}(G)$.

(ii) Consider the homomorphism $\beta : Z[X_1, X_2, \dots X_k] \to Z(G)$ given by $X_i \mapsto a_i - 1$ $(1 \leq i \leq k)$. Clearly its kernel contains the ideal $\mathfrak{b}(G)$. Since the map $a_i \mapsto (X_i + 1) + \mathfrak{b}(G)$ can be extended to a homomorphism $Z(G) \to Z[X_1, X_2, \dots, X_k]/\mathfrak{b}(G)$, β induces an isomorphism $Z[X_1, X_2, \dots, X_k]/\mathfrak{b}(G) \to Z(G)$. Under this isomorphism $\Delta_Z(G)$ corresponds to $\mathfrak{X}/\mathfrak{b}(G)$, where $\mathfrak{X}$ = the ideal of $Z[X_1, X_2, \dots, X_k]$ generated by $X_1, X_2, \dots, X_k$. Thus $\mathfrak{G}_Z(G) \cong \sum_{n \geq 0} \mathfrak{X}^n + \mathfrak{b}(G)/\mathfrak{X}^{n+1} + \mathfrak{b}(G) \cong Z[X_1, X_2, \dots, X_k]/\mathfrak{a}(G)$.

It is clear that the ideal $\mathfrak{s}(G)$ is always contained in the ideal $\mathfrak{a}(G)$. We identify the algebras $\mathfrak{s}(G)$ and $\mathfrak{G}_Z(G)$ with $Z[X_1, X_2, \dots, X_k]/\mathfrak{s}(G)$ and $Z[X_1, X_2, \dots, X_k]/\mathfrak{a}(G)$ respectively. Then the epimorphism $\theta : \mathfrak{S}(G) \to \mathfrak{G}_Z(G)$ gets identified with the natural epimorphism

(6.8) $$\theta^* : Z[X_1,X_2,\ldots,X_k]/\mathfrak{S}(G) \to Z[X_1,X_2,\ldots,X_k]/\mathfrak{A}(G).$$

Note that if $k = 1$, then θ^* is an isomorphism. For, in this case it is trivial to see that $\mathfrak{S}(G) = \mathfrak{A}(G)$.

When we restrict ourselves to finite Abelian p-groups, then the simplest case, of course, is that of elementary Abelian p-groups. Both $P_n(G)$ and $Q_n(G)$ have been computed in [62] in this case. The structure of $\mathfrak{G}_Z(G)$ is given by the following

6.9 Theorem [74]. If G is an elementary Abelian p-group of rank k, then

$$\mathfrak{G}_Z(G) \cong Z[X_1,X_2,\ldots,X_k]/\langle pX_i, X_j^pX_\ell - X_\ell^pX_j\rangle$$

Proof. We assert that the kernel of θ^* (6.8) consists of the ideal generated by $X_j^pX_\ell - X_\ell^pX_j + \mathfrak{S}(G)$ $(1 \leq j,\ell \leq k)$, or in other words the ideal $\mathfrak{A}(G)$ is generated by pX_i, $X_j^pX_\ell - X_\ell^pX_j$ $(1 \leq i,j,\ell \leq k)$. To see that $X_j^pX_\ell - X_\ell^pX_j + \mathfrak{S}(G)$ is in the kernel, we need two simple lemmas.

6.10 Lemma. Let $H = \langle a\rangle$ be a cyclic group of order p, p prime Then

$$(a-1)^p \equiv -p(a-1) \pmod{\Delta_Z^{p+1}(H)}.$$

Proof. The equation $a^p = 1$ gives

$$p(a-1) + \binom{p}{2}(a-1)^2 + \ldots + (a-1)^p = 0$$

which shows that $p\Delta_Z(H) \subseteq p\Delta_Z^2(H) + \Delta_Z^p(H)$. Hence, iteration gives $p\Delta_Z(H) \subseteq \Delta_Z^p(H)$. Thus the terms $\binom{p}{i}(a-1)^i$, $2 \leq i \leq p-1$, all belong to $\Delta_Z^{p+1}(H)$ and we have the assertion.

6.11 Lemma. Let $K = \langle a,b\rangle$ be an Abelian group, order of $a = p =$ order of b, p prime. Then

$$(a-1)^p(b-1) \equiv (b-1)^p(a-1) \pmod{\Delta_Z^{p+2}(K)}.$$

Proof. By Lemma 6.10, $(a-1)^p(b-1) \equiv -p(a-1)(b-1) \pmod{\Delta_Z^{p+2}(K)}$

and $(b-1)^p(a-1) \equiv -p(a-1)(b-1) \pmod{\Delta_Z^{p+2}(K)}$.

Hence $(a-1)^p(b-1) \equiv (b-1)^p(a-1) \pmod{\Delta_Z^{p+2}(K)}$.

and the Lemma is proved.

It follows from Lemma 6.11 that if a_i and a_j are two generators of the elementary Abelian group G, then

$$\underbrace{a_i \hat{\otimes} a_i \hat{\otimes} \cdots \hat{\otimes} a_i}_{p \text{ times}} \hat{\otimes} a_j - \underbrace{a_j \hat{\otimes} a_j \hat{\otimes} \cdots \hat{\otimes} a_j}_{p \text{ times}} \hat{\otimes} a_i$$

is in Ker θ. Hence

$$X_i^p X_j - X_j^p X_i + \mathfrak{A}(G) \qquad (1 \leqslant i,j \leqslant k)$$

are all in the kernel of θ^*. Let M_n = the additive group of the forms of degree n in $Z[X_1, X_2, \ldots, X_k]$. Let $\bar{M}_n = M_n + \mathfrak{A}(G)/\mathfrak{A}(G)$. A simple counting argument shows that for $n \geqslant (p-1)(k-1)+1$, the number of monomials $X_r^{i_r} \ldots X_k^{i_k}$ of degree n, where $1 \leqslant r \leqslant k$, $i_r > 0$ and $0 \leqslant i_s \leqslant p-1$ for $r+1 \leqslant s \leqslant k$, is $(p^k-1)/(p-1)$. Since these monomials generate $\bar{M}_n$ ($X_j^p X_\ell - X_\ell^p X_j$ being in $\mathfrak{A}(G)$), the rank of $\bar{M}_n \leqslant (p^k-1)/(p-1)$ for $n \geqslant (k-1)(p-1)+1$. Suppose $h \in \mathfrak{A}(G)$ is a form of degree s which is not in $\langle pX_i, X_j^p X - X^p X_j \rangle$. Then it is not hard to see that there will be a form of degree s of the type

$$g = X_r \Sigma c_{i_r \ldots i_k} X_r^{i_r} \ldots X_k^{i_k} \quad \text{with} \quad 0 \leqslant i_{r+j} \leqslant p-1 \quad \text{for} \quad j \geqslant 1$$

which is in $\mathfrak{A}(G)$ but not in $\langle pX_i, X_j^p X_\ell - X_\ell^p X_j \rangle$. Multiplying g by a suitable power of X_r, we get a form of degree $\geqslant (k-1)(p-1)+1$ which is in $\mathfrak{A}(G)$ showing that the monomials $X_r^{i_r} \ldots X_k^{i_k}$ of degree n in which $i_r > 0$ and $0 \leqslant i_{r+j} \leqslant p-1$ for $j \geqslant 1$ are linearly dependent in $\bar{M}_n$. This implies that the rank of $\bar{M}_n < (p^k-1)/(p-1)$ for all large n. Now $\bar{M}_n = Q_n(G)$. Thus to complete the proof of Theorem 6.9 it is enough to prove the following

6.12 Lemma. If G is an elementary Abelian p-group or rank k, then the rank of $Q_n(G) \geqslant (p^k-1)/(p-1)$ for all n sufficiently large.

This result is proved in [74]. We propose to give a slightly different proof which we postpone until the next Section.

An interesting consequence of Theorem 6.9 is the following

6.13 Corollary [62]. If G is an elementary Abelian p-group of rank k, then for $n \geqslant (p-1)(k-1)+1$, $Q_n(G)$ is elementary Abelian p-group of rank $(p^k-1)/(p-1)$.

Corollary 6.13 was proved in [62] by a different method. In fact, in [62] we first computed $P_n(G)$ for an elementary Abelian p-group and then deduced the assertion about $Q_n(G)$. This result (6.13) has led Bachmann-Grünenfelder to nice qualitative information about the groups $Q_n(G)$. They have shown, in particular, that for any finite Abelian group G, the groups $Q_n(G)$ become isomorphic to each other for sufficiently large n. We discuss this behaviour of $Q_n(G)$ in the next Section.

7. THE PERIODICITY OF THE GROUPS $Q_n(G)$

Let G be a finite group of order r, say. Then $r\Delta_Z(G) \subseteq \Delta_Z^2(G)$. Therefore, the groups $P_n(G)$ are all of finite exponent. Since these groups are finitely generated, they are all finite. Therefore, for all $m \geq 1$, the rank of the free Abelian group $\Delta_Z^m(G)$ is equal to that of $\Delta_Z(G)$ which is r-1. Now, as observed by Sandling [81], the Jordan-Zassenhaus Theorem [14] adapted to the case of finite groups implies that the set of G-modules $\{\Delta_Z^n(G)\}_{n\geq 1}$ splits into a finite number of isomorphism classes. This implies that there exist integers n_o and π such that

$$\Delta_Z^{n_o}(G) \cong \Delta_Z^{n_o+\pi}(G)$$

as G-modules. The fact that this isomorphism is that of G-modules shows that it induces a G-module isomorphism

$$\Delta_Z^{n_o+1}(G) \cong \Delta_Z^{n_o+\pi+1}(G)$$

Thus, by iteration, $\Delta_Z^n(G) \cong \Delta_Z^{n+\pi}(G)$ for all $n \geq n_o$ and we have

7.1 Theorem [5]. If G is a finite group, then there exist integers n_o and π such that

$$Q_n(G) \cong Q_{n+\pi}(G) \text{ for all } n \geq n_o.$$

Bachmann-Grünenfelder have shown that if G is a finite nilpotent group of class c, say, then the period of Theorem 7.1 divides the least common multiple of 1,2,...,c. Thus, in particular, for Abelian groups we have

7.2 Theorem ([5],[23],[88]). If G is a finite Abelian group, then the sequence $\{Q_n(G)\}$ becomes stationary (up to isomorphism) after a

finite number of steps.
(For order of the stationary value of $Q_n(G)$ see [89]).

We give here a proof of Theorem 7.2 by specializing the proof of Bachmann-Grünenfelder [5] to the Abelian case.

7.3 Definition If G is a finite Abelian p-group of order p^r, then we call r the length of G and denote it by $\ell(G)$.

Let G be a finite Abelian p-group. Let α be an integer ≥ 1. Then $Z/p^\alpha Z \otimes_Z \mathfrak{S}(G)$ is graded $Z/p^\alpha Z$-algebra ($\mathfrak{S}(G)$ = the symmetric algebra of the Abelian group G). If $M = \sum_{n\geq 0} M_n$ is a finitely generated graded module over $Z/p^\alpha Z \otimes_Z \mathfrak{S}(G)$, then each M_n is a finitely generated $Z/p^\alpha Z$-module and so is a finite Abelian p-group. We can thus define the length $\ell(M_n)$ of M_n for all $n \geq 0$. We define the Poincaré series of M to be the formal power series

$$P(M,t) = \sum_{n \geq 0} \ell(M_n) t^n$$

where t is an indeterminate.

7.4 Lemma. For each finitely generated graded module M over $Z/p^\alpha Z \otimes_Z \mathfrak{S}(G)$, the Poincaré series $P(M,t)$ is a rational function of the form $f(t)/(1-t)^k$ where k is the minimal number of generators of G and $f(t) \in Z[t]$.

Proof. We proceed by induction on k, the minimal number of generators of G. If $k = 0$, i.e. $G = 1$, then $\mathfrak{S}(G) = Z$ and $M_n = 0$ for all large n. Thus $P(M,t) \in Z[t]$. Suppose the result holds for groups having less than k generators ($k > 0$). Let $x_1, x_2, \ldots, x_k$ be a set of generators of G. Then the action of $1 \otimes x_k$ on M is a module homomorphism $M \to M$ (of graded modules) of degree 1. Let $K = \{m \in M \mid m(1\otimes x_k) = 0\}$ and $C = M/M(1\otimes x_k)$ be the kernel and cokernel of this homomorphism. Then both K and C are finitely generated graded modules over $Z/p^\alpha Z \otimes_Z \mathfrak{S}(G/\langle x_k\rangle)$ (observe that $Z/p^\alpha Z \otimes_Z \mathfrak{S}(G)$ is a commutative Noetherian ring). Therefore, induction applies to both $P(C,t)$ and $P(K,t)$ and we have $P(C,t) = f(t)/(1-t)^{k-1}$ and $P(K,t) = g(t)/(1-t)^{k-1}$ for some $f(t)$, $g(t) \in Z[t]$. Now the exact sequence

$$0 \longrightarrow K_n \longrightarrow M_n \xrightarrow{1\otimes x_k} M_{n+1} \longrightarrow C_{n+1} \longrightarrow 0$$

implies that

$$\ell(K_n) - \ell(M_n) + \ell(M_{n+1}) - \ell(C_{n+1}) = 0 \quad \text{for all } n \geq 0.$$

Multiplying by t^{n+1} and summing over n, we get

$$(1-t)P(M,t) = P(C,t) - tP(K,t) + a, \text{ for some } a \in Z.$$

Substituting for $P(C,t)$ and $P(K,t)$, we see that $P(M,t) = F(t)/(1-t)^k$ with $F(t)$ in $Z[t]$. This completes induction and the Lemma is proved.

7.5 Proof of Theorem 7.2. The natural epimorphism

$$Z/p^\alpha Z \otimes_Z \mathfrak{S}(G) \longrightarrow Z/p^\alpha Z \otimes_Z \mathfrak{G}_Z(G)$$

induced by the homomorphism (6.6) enables us to regard $Z/p^\alpha Z \otimes_Z \mathfrak{G}_Z(G)$ as a $Z/p^\alpha Z \otimes_Z \mathfrak{S}(G)$-module which is finitely generated because G is finite. Thus, by Lemma 7.4, the Poincaré series $P(Z/p^\alpha Z \otimes_Z \mathfrak{G}_Z(G),t) = F(t)/(1-t)^k$ for some $F(t) \in Z[t]$. On the other hand, Theorem 7.1 shows that there exist n_o and π such that

$$\ell(Z/p^\alpha Z \otimes_Z Q_n(G)) = \ell(Z/p^\alpha Z \otimes_Z Q_{n+\pi}(G))$$

For all $n \geq n_o$. Hence the Poincaré series must be of the form

$$f(t) + g(t)t^{n_o}/(1-t^\pi)$$

for some $f(t)$, $g(t) \in Z[t]$. Elementary unique factorization considerations show that

$$f(t) + g(t)t^{n_o}/(1-t^\pi) = F(t)/(1-t)^k$$

is possible with $f(t)$, $g(t)$, $F(t) \in Z[t]$ only if $F(t) = (1-t)^{k-1}h(t)$ for some $h(t) \in Z[t]$. This shows that for every $\alpha \geq 1$

$$P(Z/p^\alpha Z \otimes_Z \mathfrak{G}_Z(G),t) = h(t)/(1-t)$$

for some $h(t) \in Z[t]$. Thus there exists an integer N_α such that

$$\ell(Z/p^\alpha Z \otimes_Z Q_n(G)) = \ell(Z/p^\alpha Z \otimes_Z Q_{n+1}(G))$$

for all $n \geq N_\alpha$.

Suppose G has exponent p^e. Then $p^e\Delta_Z(G) \subseteq \Delta_Z^2(G)$ and therefore $p^e\Delta_Z^n(G) \subseteq \Delta_Z^{n+1}(G)$ for all $n \geq 1$. Thus $Q_n(G)$ has exponent a power of $p \leq p^e$, for all $n \geq 1$. Let $s(n,r)$ denote the number of cyclic components of order p^r in a primary decomposition of $Q_n(G)$. Note that $s(n,r) = 0$ if $r > e$. Taking $N = \text{Max}(N_1,N_2,\ldots,N_e)$, we have

$$\ell(Z/p^\alpha Z \otimes_Z Q_n(G)) = \ell(Z/p^\alpha Z \otimes_Z Q_{n+1}(G)) \tag{7.6}$$

for all $e \geq \alpha \geq 1$, $n \geq N$. Now recall that $Z/p^{\alpha}Z \otimes_Z Z/p^{\beta}Z \cong Z/p^{\gamma}Z$ where $\gamma = \mathrm{Min}(\alpha,\beta)$. Taking $\alpha = 1,2,\ldots,e$ in (7.6), we get e equations

$$E_\alpha : \sum_{i=1}^{\alpha-1} is(n,i) + \alpha \sum_{i=\alpha}^{e} s(n,i) = \sum_{i=1}^{\alpha-1} is(n+1,i) + \alpha \sum_{i=\alpha}^{e} s(n+1,i).$$

Subtracting $E_{\alpha-1}$ from E_α gives

$$s(n,\alpha) = s(n+1,\alpha)$$

for all $n \geq N$. This proves that $Q_n(G) \cong Q_{n+1}(G)$ for all $n \geq N$.

7.7 <u>Remark</u>. Losey-Losey ([46], [47]) have computed the augmentation quotients of groups of order p^3 (p prime) and also for those finite p-groups G whose lower central series is a restricted N-series relative to p. Their calculations provide specific values of n_o and π in the Bachmann-Grünenfelder Theorem 7.1 in these cases.

7.8 <u>Proof of Lemma</u> 6.12. By Theorem 7.2 there exists n_o such that $\ell(Q_n(G)) = \ell(Q_{n+1}(G)) = e$, say, for all $n \geq n_o$. From the exact sequence

$$0 \to Q_n(G) \to P_n(G) \to P_{n-1}(G) \to 0 \qquad (n \geq 1)$$

it follows that

$$\ell(P_n(G)) = en+f$$

for all $n \geq n_o$, where $f = \ell(P_{n_o}(G)) - en_o$. Now we can show that

$$\Delta_Z^{(n+k-1)(p-1)+1}(G) \subseteq p^n\Delta_Z(G) \quad \text{for all } n \geq 1$$ (see [81] for a more general result).

Hence
$$n(p^k-1) = \ell(\Delta_Z(G)/p^n\Delta_Z(G)) \leq \ell(P_{(n+k-1)(p-1)}(G))$$
$$= e(n+k-1)(p-1) + f \quad \text{for all large } n.$$

This is possible only if $e \geq (p^k-1)/(p-1)$. As $Q_n(G)$ is an elementary Abelian p-group, the rank of $Q_n(G)$ is equal to the length of $Q_n(G)$. Hence the rank of $Q_n(G) \geq (p^k-1)/(p-1)$ for all large n.

8. <u>THE HOMOMORPHISM</u> θ_n

In this Section we examine the natural epimorphism

$$\theta_n : SP^n(G) \to Q_n(G), \quad x_1 \hat{\otimes} x_2 \hat{\otimes} \ldots \hat{\otimes} x_n \longmapsto (x_1-1)(x_2-1)\ldots(x_n-1) + \Delta_Z^{n+1}(G)$$

for Abelian groups G.

Consider the map

$$\varphi_n : G \to SP^n(G)$$

given by

$$\varphi_n(x) = \underbrace{x \hat{\otimes} x \hat{\otimes} \cdots \hat{\otimes} x}_{n \text{ times}}, \quad n \geq 1 .$$

We have seen in Chapter V that φ_n is a polynomial map of degree $\leq n$. Thus φ_n induces a homomorphism

$$\varphi_n : Q_n(G) \to SP^n(G) .$$

The calculations of Chapter V give the following

8.1 Theorem [98]. For all $n \geq 1$

$$\varphi_n \circ \theta_n = n!$$

and

$$\theta_n \circ \varphi_n = n!$$

8.2 Corollary ([3], [64]). If G is a torsion-free Abelian group, then

$$\theta : \mathfrak{S}(G) \to \mathfrak{G}_Z(G)$$

is an isomorphism.

Proof. It suffices to prove that each θ_n is an isomorphism. Now if G is a torsion-free Abelian groups, then so is $SP^n(G)$ for all $n \geq 1$. This can be seen by noting that it suffices to consider finitely generated groups. If G is a finitely generated torsion-free Abelian group, then by Proposition 6.7 $\mathfrak{S}(G)$ is isomorphic to a ring of polynomials over Z and so is torsion-free as an additive group. Thus the Corollary is an immediate consequence of Theorem 8.1.

8.3 Remark. The homomorphism $\theta : \mathfrak{S}(G) \to \mathfrak{G}_Z(G)$ is not always an isomorphism. For example, as we have seen, if G is an elementary Abelian p-group of order p^m, $m \geq 2$, then the rank of $Q_n(G)$ becomes stationary for all large n while that of $SP^n(G)$ keeps on increasing with n (Corollary 6.13 and Proposition 6.7). Recently Passi-Vermani [74] have calculated the exact stage upto which θ_n is an isomorphism.

8.4 Theorem [74]. Let $G = Z_{p^{m_1}} \oplus Z_{p^{m_2}} \oplus \cdots \oplus Z_{p^{m_k}}$, $k > 1$, $r = \operatorname{Min}_{1 \leq i < j \leq k} |m_i - m_j|$. Then θ_n is an isomorphism if and only if $n \leq p + r(p-1)$.

The proof of Theorem 8.4 is rather technical and is omitted. It is achieved by the methods illustrated in Section 6. Note that it follows from Theorems 6.3 and 8.4 that if G is a finite Abelian group and $\theta : \mathfrak{S}(G) \to \mathfrak{G}_Z(G)$ is an isomorphism, then G must be cyclic. We have already noted in Section 6 that if G is cyclic, then θ is an isomorphism. Thus we have

8.5 *Corollary* [3]. *Let* G *be a finite Abelian group*. *Then* $\theta : \mathfrak{S}(G) \to \mathfrak{G}_Z(G)$ *is an isomorphism if and only if* G *is cyclic*.

It follows from Theorem 8.4 that if G is a finite Abelian p-group then $\theta_2 : SP^2(G) \to Q_2(G)$ is an isomorphism. From this fact it can be deduced that θ_2 is always an isomorphism (use reduction Theorems 6.3 and 6.4). This can, however, be proved directly without having to go through the reductions.

8.6 *Theorem* [64]. *For every Abelian group* G ,

$$\theta_2 : SP^2(G) \to Q_2(G)$$

is an isomorphism.

Proof. Let T = the additive group of rationals mod 1. Let $\alpha : SP^2(G) \to T$ be a homomorphism. Then the map $f : G \times G \to T$ given by $f(x,y) = \alpha(x \hat{\otimes} y)$ is a symmetric bilinear map. Therefore, f is a symmetric 2-cocycle on G . Hence f must be a coboundary. For, the extension

$$0 \to T \to M \to G \to 0$$

of T by G defined by f is Abelian (because f is symmetric) and therefore splits (because T is divisible Abelian). Let $\chi : G \to T$ be a map such that $\chi(1) = 0$ and $f(x,y) = \chi(xy) - \chi(x) - \chi(y)$, $x,y \in G$. The bilinearity of f implies that χ is a polynomial map of degree ≤ 2 . Hence χ induces a homomorphism $\chi^* : Z(G)/\Delta_Z^3(G) \to T$. By definitions, we have $\alpha = \chi^* \circ \theta_2$. It follows that α vanishes on $\operatorname{Ker} \theta_2$. By the divisibility of T it follows that every homomorphism $\operatorname{Ker} \theta_2 \to T$ is zero. Hence $\operatorname{Ker} \theta_2 = 0$. As θ_2 is always an epimorphism, the Theorem is proved.

Using Theorem 8.6 it is easy to compute $Q_2(G)$ in general.

Let $\mathfrak{U}_2(LG)$ = the second homogeneous component in the universal envelope of the Lie ring $LG = \sum_{i \geq 1} \gamma_i(G)/\gamma_{i+1}(G)$.

$$\mathfrak{U}_2(LG) = \{G/\gamma_2(G) \otimes G/\gamma_2(G) \oplus \gamma_2(G)/\gamma_3(G)\}/R ,$$

where R is the subgroup generated by the elements of the type

$$\tilde{x}\otimes\tilde{y} - \tilde{y}\otimes\tilde{x} - \widetilde{(x,y)} \qquad (\tilde{x},\tilde{y} \in G/\gamma_2(G)) .$$

8.7 Theorem [4]. *For every group* G,

$$Q_2(G) \cong \mathfrak{u}_2(LG) .$$

Proof. We have a commutative diagram

$$\begin{array}{ccccccccc} 0 \longrightarrow & \gamma_2(G)/\gamma_3(G) & \xrightarrow{i^*} & \mathfrak{u}_2(LG) & \xrightarrow{j^*} & \mathfrak{u}_2(L(G/\gamma_2(G))) & \longrightarrow 0 \\ & \| & & \downarrow \theta_2(G) & & \downarrow \theta_2(G/\gamma_2(G)) & \\ 0 \longrightarrow & \gamma_2(G)/\gamma_3(G) & \xrightarrow{i} & Q_2(G) & \xrightarrow{j} & Q_2(G/\gamma_2(G)) & \longrightarrow 0 \end{array}$$

with exact rows in which the homomorphisms are defined as follows: for $x \in \gamma_2(G)$, $i^*(x\gamma_3(G)) = x\gamma_3(G) + R$, $i(x\gamma_3(G)) = x-1+ \Delta_Z^3(G)$, j^* and j are the homomorphisms induced by the natural projection $G \to G/\gamma_2(G)$. Since $\theta_2(G/\gamma_2(G))$ is an isomorphism (Theorem 8.6), it follows that $\theta_2(G)$ is an isomorphism.

The natural short exact sequence

(8.8) $$0 \to \gamma_2(G)/\gamma_3(G) \xrightarrow{i} Q_2(G) \xrightarrow{j} Q_2(G/\gamma_2(G)) \to 0$$

has been studied by several authors ([4], [24], [44], [70], [72], [80]). The point of interest here has been whether or not this sequence splits.

Sandling [80] showed that this sequence splits if G is finite, and Losey [44] did the same for G finitely generated (see also [4] and [72]).

Let $A = G/\gamma_2(G)$. Then we have a commutative diagram

(8.9) $$\begin{array}{ccccccc} 0 \to & C & \to & A\otimes A & \xrightarrow{u} & SP^2(A) & \to 0 \\ & \downarrow & & \downarrow \beta & & \downarrow \theta_2(A) & \\ 0 \to & \gamma_2(G)/\gamma_3(G) & \to & Q_2(G) & \to & Q_2(A) & \to 0 \end{array}$$

where C is the subgroup $\langle x\otimes y - y\otimes x \mid x,y \in A\rangle$ of $A\otimes A$, with exact rows, u is the natural projection from the tensor product to the symmetric product and $\beta(\tilde{x}\otimes\tilde{y}) = (x-1)(y-1) + \Delta_Z^3(G)$, $\tilde{z} = z\gamma_2(G)$ for $z \in G$. It is clear that if the top row in (8.9) splits, then so does the bottom row. Thus sufficient conditions for the splitting of (8.8) can be obtained by studying the exact sequence

(8.10) $$0 \to C \to A\otimes A \to SP^2(A) \to 0$$

for Abelian groups A. This is the point of view taken in [24]. Some of the cases in which it is trivial to see that (8.10) splits are:

(i) A locally cyclic (here $C = 0$) ;

(ii) A uniquely 2-divisible (here $\langle x \otimes x \mid x \in A \rangle$ is a complement to C in $A \otimes A$).

8.11 Proposition [24]. Let $A = \sum_{i \in I} A_i$ be an Abelian group and suppose that, for each i, the subgroup $C_i = \langle x \otimes y - y \otimes x \mid x,y \in A_i \rangle$ of $A_i \otimes A_i$ is a summand of $A_i \otimes A_i$. Then $C = \langle x \otimes y - y \otimes x \mid x,y \in A \rangle$ is a summand of $A \otimes A$.

Proof. For each i, let B_i be a complement to C_i in $A_i \otimes A_i$. Impose a fixed linear order on I. Then $A \otimes A = \sum_i A_i \otimes A_i + \sum_{i \neq j} A_i \otimes A_j$ and $B = \sum_i B_i \oplus \sum_{i<j} A_i \otimes A_j$ is a complement of C in $A \otimes A$.

Thus we have

8.12 Theorem [24]. The sequence (8.8) splits if $G/\gamma_2(G)$ is a direct sum of cyclic groups, or is divisible, or is completely decomposable torsion-free (i.e. a direct sum of rank one groups).

8.13 Corollary.

If G is finitely generated, then

$$Q_2(G) \cong SP^2(G/\gamma_2(G)) \oplus \gamma_2(G)/\gamma_3(G) .$$

8.14 Remarks.

(i) It has been shown in [24] that the sequence (8.8) splits when $G/\gamma_2(G)$ is a torsion Abelian group. Also in [24] a counter-example is given to show that the sequence (8.8) does not split in general.

(ii) The splitting of (8.8) is related to the problem of realizing a group G of class 2 as the circle group $(R,\circ)$ of a nilpotent ring R (i.e. R regarded as a group under the 'circle operation': $a \circ b = a+b+ab$, $a,b \in R$) of index 3 (i.e. $R^3 = 0$) ([2], [24], [83]). It is shown in [24] that groups of class 2 cannot be realized as circle groups of nilpotent rings of index 3 in general.

(iii) Tahara ([95], [97]) has computed the structures of $Q_3(G)$ and $Q_4(G)$ for finite groups G.

REFERENCES

[1] ATIYAH, M.F. and MACDONALD, I.G., Introduction to Commutative Algebra, Addison-Wesley, 1969.

[2] AULT, J.C. and WATTERS, J.F., Circle groups of nilpotent rings, Amer. Math. Monthly, 80 (1973), 48-52.

[3] BACHMANN, F. and GRUENENFELDER, L., Ueber Lie-Ringe von Gruppen und ihre universellen Enveloppen, Comment. Math. Helv. 47 (1972), 332-340.

[4] BACHMANN, F. and GRUENENFELDER, L., Homological methods and the third dimension subgroup, Comment. Math. Helv. 47 (1972), 526-531.

[5] BACHMANN, F. and GRUENENFELDER, L., The periodicity in the graded ring associated with an integral group ring, J. Pure Appl. Algebra 5 (1974), 253-264.

[6] BOVDI, A.A., Intersections of powers of the fundamental ideal of an integral group ring, Mat. Zametki 2 (1967), 129-132.

[7] BOVDI, A.A., Dimension subgroups, Proceedings of the Riga Seminar on Algebra, Latv. Gos. Univ., Riga, 1969, 5-7.

[8] BUCKLEY, J., Polynomial functions and wreath products, Ill. J. Math. 14 (1970), 274-282.

[9] CHEN, K.T., A group ring method for finitely generated groups, Trans. Amer. Math. Soc. 76 (1954), 275-287.

[10] CHEN, K.T., FOX, R.H. and LYNDON, R.C., Free differential calculus IV: The quotient groups of the lower central series, Ann. of Math. 68 (1958), 81-95.

[11] COHN, P.M., Generalization of a theorem of Magnus, Proc. London Math. Soc. (3) 2 (1952), 297-310.

[12] CONNELL, I.G., On the group ring, Canad. J. Math. 15 (1963), 650-685.

[13] CROWELL, R.H. and FOX, R.H., Introduction to Knot Theory, Ginn and Co., Boston, 1963.

[14] CURTIS, C. and REINER, I., Representation Theory of Finite Groups and Associative Algebras, Interscience, New York, 1962.

[15] DARK, R., On Nilpotent Products of Groups of Prime Order, Thesis, Cambridge University, Cambridge, 1968.

[16] DYNKIN, E.B., Normed Lie algebras and analytic groups, Uspehi Mat. Nauk (N.S.) 5 (1950), 135-186 (Russian); Amer. Math. Soc. Translations 97, Providence, R.I., 1953.

[17] EILENBERG, S. and MAC LANE, S., On the groups $H(\Pi,n)$-II: Methods of Computation, Ann. of Math. 60 (1954), 49-139.

[18] FORMANEK, E., A short proof of a theorem of Jennings, Proc. Amer. Math. Soc. 26 (1970), 405-407.

[19] GRUENBERG, K.W., Residual properties of infinite soluble groups, Proc. London Math. Soc. (3) 7 (1957), 29-62.

[20] GRUENBERG, K.W., The residual nilpotence of certain presentations of finite groups, Arch. Math. 13 (1962), 408-417.

[21] GRUENBERG, K.W. and ROSEBLADE, J.E., The augmentation terminals of certain locally finite groups, Canad. J. Math. 24 (1972), 221-238.

[22] GUPTA, N.D. and PASSI, I.B.S., Some properties of Fox subgroups of free groups, J. Algebra 43 (1976), 198-211.

[23] GUTIERREZ, M.A., The $\bar{\mu}$-invariants for groups, Proc. Amer. Math. Soc., 55 (1976), 293-298.

[24] HALES, A.W. and PASSI, I.B.S., The second augmentation quotient of an integral group ring, Arch. Math. (to appear).

[25] HALL, M.Jr., The Theory of Groups, Macmillan, New York, 1959.

[26] HALL, P., Nilpotent Groups, Canadian Mathematical Congress, University of Alberta, Edmonton, 1957; Queen Mary College Math. Notes, 1970.

[27] HARTLEY, B., The residual nilpotence of wreath products, Proc. London Math. Soc. (3) 20 (1970), 365-392.

[28] HARTLEY, B., On residually finite p-groups, Symposia Mathematica 17 (1976), 225-234.

[29] HARTLEY, B. and HAWKES, T.O., Rings, Modules and Linear Algebra, Chapman and Hall, 1970.

[30] HILTON, P.J. and STAMMBACH, U., A Course in Homological Algebra, Springer-Verlag, New York, Heidelberg, Berlin, 1970.

[31] HOARE, A.H.M., Group rings and lower central series, J. London Math. Soc. (2) 1 (1969), 37-40.

[32] HUPPERT, B., Endliche Gruppen-I, Springer-Verlag, Berlin, Heidelberg, New York, 1967.

[33] HURLEY, T.C., Residual properties of groups determined by ideals, Proc. Roy. Irish Acad. 77 A (1977), 97-104.

[34] JACOBSON, N., Lie Algebras, Interscience, New York, 1962.

[35] JENNINGS, S.A., The structure of the group ring of a p-group over a modular field, Trans. Amer. Math. Soc. 50 (1941), 175-185.

[36] JENNINGS, S.A., The group ring of a class of infinite nilpotent groups, Canad. J. Math. 7 (1955), 169-187.

[37] KAPLANSKY, I., Infinite Abelian Groups, The University of Michigan Press, Ann Arbor, 1954.

[38] LAZARD, M., Sur les groupes nilpotents et les anneaux de Lie, Ann. École Norm. Sup. 71 (1954), 101-190.

[39] LICHTMAN, A.I., The residual nilpotence of the augmentation ideal and the residual nilpotence of some classes of groups, Israel J. Math. 26 (1977), 276-293.

[40] LICHTMAN, A.I., Necessary and sufficient conditions for the residual nilpotence of free products of groups, J. Pure and Appl. Algebra 12 (1978), 49-64.

[41] LOMBARDO-RADICE, L., Interno alle algebre legate ai gruppi di ordine finito - II, Rend. Sem. Mat. Roma 3 (1939), 239-256.

[42] LOSEY, G., On dimension subgroups, Trans. Amer. Math. Soc. 97 (1960), 474-486.

[43] LOSEY, G., On group algebras of p-groups, Mich. Math. J. 7(1960), 237-240.

[44] LOSEY, G., On the structure of $Q_2(G)$ for finitely generated groups, Canad. J. Math. 25 (1973), 353-359.

[45] LOSEY, G., N-series and filtrations of the augmentation ideal, Canad. J. Math. 26 (1974), 962-977.

[46] LOSEY, G. and LOSEY, N., Augmentation quotients of some non-Abelian groups (to appear).

[47] LOSEY, G. and LOSEY, N., The stable behaviour of the augmentation quotients of the groups of order p^3 (to appear).

[48] MAGNUS, W., Ueber Beziehungen zwischen höheren Kommutatoren, J. Reine Angew. Math. 177 (1937), 105-115.

[49] MAGNUS, W., On a theorem of Marshall Hall, Ann. of Math. 40(1939), 764-768.

[50] MAL'CEV, A.I., Generalized nilpotent algebras and their adjoint groups, Mat. Sbornik N.S. 25(67) (1949), 347-366; Amer. Math. Soc. Transl. (2) 69 (1968), 1-21.

[51] MELDRUM, J.D.P., Central series in wreath products, Proc. Cambridge Philos. Soc. 63(1967), 551-567.

[52] MELDRUM, J.D.P., On nilpotent wreath products, Proc. Cambridge Philos Soc. 68 (1970), 1-15.

[53] MELDRUM, J.D.P., Correction to a paper on wreath products, Proc. Cambridge Philos. Soc. 76 (1974), 21.

[54] MILNOR, J. and MOORE, J., On the structure of Hopf algebras, Ann. of Math. 81 (1965), 211-264.

[55] MITAL, J.N., On residual nilpotence, J. London Math. Soc. (2)2 (1970), 337-345.

[56] MORAN, S., Dimension subgroups mod n , Proc. Cambridge Philos. Soc. 68 (1970), 579-582.

[57] PARMENTER, M.M., On a theorem of Bovdi, Canad. J. Math. 23(1971), 929-932.

[58] PARMENTER, M.M. and PASSI, I.B.S., The intersection theorem in group rings (unpublished).

[59] PARMENTER, M.M., PASSI, I.B.S. and SEHGAL, S.K., Polynomial ideals in group rings, Canad. J. Math. 25 (1973), 1174-1182.

[60] PARMENTER, M.M., PASSI, I.B.S. and SEHGAL, S.K., The nilpotent residue of the augmentation ideal of a group ring (unpublished).

[61] PARMENTER, M.M. and Sehgal, S.K., Idempotent elements and ideals in group rings and the intersection theorem, Arch. Math. 24 (1973), 586-600.

[62] PASSI, I.B.S., Polynomial maps on groups, J. Algebra 9(1968), 121-151.

[63] PASSI, I.B.S., Dimension subgroups. J. Algebra 9(1968), 152-182.

[64] PASSI, I.B.S., Polynomial functors, Proc. Cambridge Philos. Soc. 66(1969), 505-512.

[65] PASSI, I.B.S., Induced central extensions, J. Algebra 16(1970), 27-39.

[66] PASSI, I.B.S., Polynomial maps, Proc. Second Internat. Conf. Theory of Groups, Canberra, 1973, Lecture Notes in Math. Vol. 372(1974), 550-561, Springer-Verlag, Berlin, Heidelberg, New York.

[67] PASSI, I.B.S., Polynomial maps on groups-II, Math. Z. 135(1974), 137-141.

[68] PASSI, I.B.S., Annihilators of relation modules-II, J. Pure and Appl. Algebra 6(1975), 235-237.

[69] PASSI, I.B.S., The associated graded ring of a group ring, Bull. London Math. Soc. (to appear).

[70] PASSI, I.B.S. and SEHGAL, S.K., Isomorphism of modular group algebras, Math. Z. 129 (1972), 65-73.

[71] PASSI, I.B.S. and SEHGAL, S.K., Lie dimension subgroups, Comm. Algebra 3 (1975), 59-73.

[72] PASSI, I.B.S. and SHARMA, S., The third dimension subgroup mod n, J. London Math. Soc. (2) 9 (1974), 176-182.

[73] PASSI, I.B.S. and STAMMBACH, U., A filtration of Schur multiplicator, Math. Z. 135 (1974), 143-148.

[74] PASSI, I.B.S. and VERMANI, L.R., The associated graded ring of an integral group ring, Proc. Cambridge Philos. Soc. 82 (1977), 25-33.

[75] PASSMAN, D.S., The Algebraic Structure of Group Rings, Interscience, New York, 1977.

[76] PLOTKIN, B.I., Remarks on stable representation of nilpotent groups, Trans. Moscow Math. Soc. 29 (1973), 185-200.

[77] QUILLEN, D., On the associated graded ring of a group ring, J. Algebra 10 (1968), 411-418.

[78] RIPS, E., On the fourth integer dimension subgroup, Israel J. Math. 12 (1972), 342-346.

[79] SANDLING, R., The dimension subgroup problem, J. Algebra 21 (1972), 216-231.

[80] SANDLING, R., Dimension subgroups over arbitrary coefficient rings, J. Algebra 21 (1972), 250-265.

[81] SANDLING, R., Modular augmentation ideals, Proc. Cambridge Philos. Soc. 71 (1972), 25-32.

[82] SANDLING, R., Subgroups dual to dimension subgroups, Proc. Cambridge Philos. Soc. 71 (1972), 33-38.

[83] SANDLING, R., Group rings of circle and unit groups, Math. Z. 140 (1974), 195-202.

[84] SANDLING, R. and TAHARA, K., Augmentation quotients of group rings and symmetric powers (to appear).

[85] SCHMIDT, B.K., Mappings of degree n from groups to Abelian groups, Thesis, Princeton University, Princeton, 1972.

[86] SHARMA, S., A bound for the degree of $H^2(G,Z_p)$, Canad. J. Math. 26(1974),1010-1015.

[87] SHIELD, D., The class of a nilpotent wreath product, Bull. Australian Math. Soc. 17(1977), 53-90.

[88] SINGER, M., On the augmentation terminal of a finite Abelian group J. Algebra 41(1976), 196-201.

[89] SINGER, M., Determination of the augmentation terminal for finite Abelian groups, Bull. Amer. Math. Soc. 83(1977), 1321-1322.

[90] SJOGREN, J.A., The dimension and lower central series, J. Pure and Appl. Algebra (to appear).

[91] SMITH, P.F., On the intersection theorem, Proc. London Math. Soc. (3)21 (1970), 385-398.

[92] STALLINGS, J.R., Quotients of the powers of the augmentation ideal in a group ring, Knots, Groups and 3-Manifolds, Ann. of Math. Studies 84 (1975), 101-118.

[93] SWAN, R.G., Representations of polycyclic groups, Proc. Amer. Math. Soc. 18 (1967), 573-574.

[94] TAHARA, K., The fourth dimension subgroups and polynomial maps, J. Algebra 45 (1977), 102-131.

[95] TAHARA, K., On the structure of $Q_3(G)$ and the fourth dimension subgroups, Japan J. Math. 3 (1977), 381-394.

[96] TAHARA, K., The fourth dimension subgroups and polynomial maps-II, Nagoya Math. J. 69 (1978), 1-7.

[97] TAHARA, K., The augmentation quotients of group rings and the fifth dimension subgroups (to appear).

[98] VERMANI, L.R., On polynomial groups, Proc. Cambridge Philos. Soc. 68 (1970), 285-289.

[99] ZASSENHAUS, H., Ein Verfahren, jeder endlichen p-Gruppe einen Lie-Ring mit der Charakteristik p zuzuordnen, Abh. Math. Sem. Hamburg 13 (1940), 200-207.

Vol. 609: General Topology and Its Relations to Modern Analysis and Algebra IV. Proceedings 1976. Edited by J. Novák. XVIII, 225 pages. 1977.

Vol. 610: G. Jensen, Higher Order Contact of Submanifolds of Homogeneous Spaces. XII, 154 pages. 1977.

Vol. 611: M. Makkai and G. E. Reyes, First Order Categorical Logic. VIII, 301 pages. 1977.

Vol. 612: E. M. Kleinberg, Infinitary Combinatorics and the Axiom of Determinateness. VIII, 150 pages. 1977.

Vol. 613: E. Behrends et al., L^P-Structure in Real Banach Spaces. X, 108 pages. 1977.

Vol. 614: H. Yanagihara, Theory of Hopf Algebras Attached to Group Schemes. VIII, 308 pages. 1977.

Vol. 615: Turbulence Seminar, Proceedings 1976/77. Edited by P. Bernard and T. Ratiu. VI, 155 pages. 1977.

Vol. 616: Abelian Group Theory, 2nd New Mexico State University Conference, 1976. Proceedings. Edited by D. Arnold, R. Hunter and E. Walker. X, 423 pages. 1977.

Vol. 617: K. J. Devlin, The Axiom of Constructibility: A Guide for the Mathematician. VIII, 96 pages. 1977.

Vol. 618: I. I. Hirschman, Jr. and D. E. Hughes, Extreme Eigen Values of Toeplitz Operators. VI, 145 pages. 1977.

Vol. 619: Set Theory and Hierarchy Theory V, Bierutowice 1976. Edited by A. Lachlan, M. Srebrny, and A. Zarach. VIII, 358 pages. 1977.

Vol. 620: H. Popp, Moduli Theory and Classification Theory of Algebraic Varieties. VIII, 189 pages. 1977.

Vol. 621: Kauffman et al., The Deficiency Index Problem. VI, 112 pages. 1977.

Vol. 622: Combinatorial Mathematics V, Melbourne 1976. Proceedings. Edited by C. Little. VIII, 213 pages. 1977.

Vol. 623: I. Erdelyi and R. Lange, Spectral Decompositions on Banach Spaces. VIII, 122 pages. 1977.

Vol. 624: Y. Guivarc'h et al., Marches Aléatoires sur les Groupes de Lie. VIII, 292 pages. 1977.

Vol. 625: J. P. Alexander et al., Odd Order Group Actions and Witt Classification of Innerproducts. IV, 202 pages. 1977.

Vol. 626: Number Theory Day, New York 1976. Proceedings. Edited by M. B. Nathanson. VI, 241 pages. 1977.

Vol. 627: Modular Functions of One Variable VI, Bonn 1976. Proceedings. Edited by J.-P. Serre and D. B. Zagier. VI, 339 pages. 1977.

Vol. 628: H. J. Baues, Obstruction Theory on the Homotopy Classification of Maps. XII, 387 pages. 1977.

Vol. 629: W. A. Coppel, Dichotomies in Stability Theory. VI, 98 pages. 1978.

Vol. 630: Numerical Analysis, Proceedings, Biennial Conference, Dundee 1977. Edited by G. A. Watson. XII, 199 pages. 1978.

Vol. 631: Numerical Treatment of Differential Equations. Proceedings 1976. Edited by R. Bulirsch, R. D. Grigorieff, and J. Schröder. X, 219 pages. 1978.

Vol. 632: J.-F. Boutot, Schéma de Picard Local. X, 165 pages. 1978.

Vol. 633: N. R. Coleff and M. E. Herrera, Les Courants Résiduels Associés à une Forme Méromorphe. X, 211 pages. 1978.

Vol. 634: H. Kurke et al., Die Approximationseigenschaft lokaler Ringe. IV, 204 Seiten. 1978.

Vol. 635: T. Y. Lam, Serre's Conjecture. XVI, 227 pages. 1978.

Vol. 636: Journées de Statistique des Processus Stochastiques, Grenoble 1977, Proceedings. Edité par Didier Dacunha-Castelle et Bernard Van Cutsem. VII, 202 pages. 1978.

Vol. 637: W. B. Jurkat, Meromorphe Differentialgleichungen. VII, 194 Seiten. 1978.

Vol. 638: P. Shanahan, The Atiyah-Singer Index Theorem, An Introduction. V, 224 pages. 1978.

Vol. 639: N. Adasch et al., Topological Vector Spaces. V, 125 pages. 1978.

Vol. 640: J. L. Dupont, Curvature and Characteristic Classes. X, 175 pages. 1978.

Vol. 641: Séminaire d'Algèbre Paul Dubreil, Proceedings Paris 1976–1977. Edité par M. P. Malliavin. IV, 367 pages. 1978.

Vol. 642: Theory and Applications of Graphs, Proceedings, Michigan 1976. Edited by Y. Alavi and D. R. Lick. XIV, 635 pages. 1978.

Vol. 643: M. Davis, Multiaxial Actions on Manifolds. VI, 141 pages. 1978.

Vol. 644: Vector Space Measures and Applications I, Proceedings 1977. Edited by R. M. Aron and S. Dineen. VIII, 451 pages. 1978.

Vol. 645: Vector Space Measures and Applications II, Proceedings 1977. Edited by R. M. Aron and S. Dineen. VIII, 218 pages. 1978.

Vol. 646: O. Tammi, Extremum Problems for Bounded Univalent Functions. VIII, 313 pages. 1978.

Vol. 647: L. J. Ratliff, Jr., Chain Conjectures in Ring Theory. VIII, 133 pages. 1978.

Vol. 648: Nonlinear Partial Differential Equations and Applications, Proceedings, Indiana 1976–1977. Edited by J. M. Chadam. VI, 206 pages. 1978.

Vol. 649: Séminaire de Probabilités XII, Proceedings, Strasbourg, 1976–1977. Edité par C. Dellacherie, P. A. Meyer et M. Weil. VIII, 805 pages. 1978.

Vol. 650: C*-Algebras and Applications to Physics. Proceedings 1977. Edited by H. Araki and R. V. Kadison. V, 192 pages. 1978.

Vol. 651: P. W. Michor, Functors and Categories of Banach Spaces. VI, 99 pages. 1978.

Vol. 652: Differential Topology, Foliations and Gelfand-Fuks-Cohomology, Proceedings 1976. Edited by P. A. Schweitzer. XIV, 252 pages. 1978.

Vol. 653: Locally Interacting Systems and Their Application in Biology. Proceedings, 1976. Edited by R. L. Dobrushin, V. I. Kryukov and A. L. Toom. XI, 202 pages. 1978.

Vol. 654: J. P. Buhler, Icosahedral Golois Representations. III, 143 pages. 1978.

Vol. 655: R. Baeza, Quadratic Forms Over Semilocal Rings. VI, 199 pages. 1978.

Vol. 656: Probability Theory on Vector Spaces. Proceedings, 1977. Edited by A. Weron. VIII, 274 pages. 1978.

Vol. 657: Geometric Applications of Homotopy Theory I, Proceedings 1977. Edited by M. G. Barratt and M. E. Mahowald. VIII, 459 pages. 1978.

Vol. 658: Geometric Applications of Homotopy Theory II, Proceedings 1977. Edited by M. G. Barratt and M. E. Mahowald. VIII, 487 pages. 1978.

Vol. 659: Bruckner, Differentiation of Real Functions. X, 247 pages. 1978.

Vol. 660: Equations aux Dérivée Partielles. Proceedings, 1977. Edité par Pham The Lai. VI, 216 pages. 1978.

Vol. 661: P. T. Johnstone, R. Paré, R. D. Rosebrugh, D. Schumacher, R. J. Wood, and G. C. Wraith, Indexed Categories and Their Applications. VII, 260 pages. 1978.

Vol. 662: Akin, The Metric Theory of Banach Manifolds. XIX, 306 pages. 1978.

Vol. 663: J. F. Berglund, H. D. Junghenn, P. Milnes, Compact Right Topological Semigroups and Generalizations of Almost Periodicity. X, 243 pages. 1978.

Vol. 664: Algebraic and Geometric Topology, Proceedings, 1977. Edited by K. C. Millett. XI, 240 pages. 1978.

Vol. 665: Journées d'Analyse Non Linéaire. Proceedings, 1977. Edité par P. Bénilan et J. Robert. VIII, 256 pages. 1978.

Vol. 666: B. Beauzamy, Espaces d'Interpolation Réels: Topologie et Géometrie. X, 104 pages. 1978.

Vol. 667: J. Gilewicz, Approximants de Padé. XIV, 511 pages. 1978.

Vol. 668: The Structure of Attractors in Dynamical Systems. Proceedings, 1977. Edited by J. C. Martin, N. G. Markley and W. Perrizo. VI, 264 pages. 1978.

Vol. 669: Higher Set Theory. Proceedings, 1977. Edited by G. H. Müller and D. S. Scott. XII, 476 pages. 1978.